ENGINEERING VISION TECHNOLOGY
Revolution and Optimism

ENGINEERING VISION TECHNOLOGY
Revolution and Optimism

Purnendu Ghosh
Executive Director
Birla Institute of Scientific Research
Jaipur, Rajasthan
Vice President
Indian National Academy of Engineering

CRC Press is an imprint of the
Taylor & Francis Group, an **informa** business

NEW INDIA PUBLISHING AGENCY
New Delhi – 110 034

First published 2021
by CRC Press
2 Park Square, Milton Park, Abingdon, Oxon, OX14 4RN

and by CRC Press
6000 Broken Sound Parkway NW, Suite 300, Boca Raton, FL 33487-2742

CRC Press is an imprint of Informa UK Limited

British Library Cataloguing-in-Publication Data
A catalogue record for this book is available from the British Library

Library of Congress Cataloging-in-Publication Data
A catalog record has been requested

ISBN: 978-1-032-01213-1 (hbk)

This book is for my students
Who are now my teachers

Preface

Writing this book was an opportunity for me to thank my teachers, my peers, my students, and my people on the occasion of completion of 50 years of my graduation in engineering.

So far it has been a wonderful journey. I am still learning. I have not forgotten how I began my journey. I have received more from my students and peers than I could give back to them. I have learnt that aesthetic development of the senses is as important, as the intellectual. I have learnt, more than the mind, it is the heart that makes a teacher. A mere intellectual teacher is a mere abstraction of facts. Teaching is truly an inward human activity.

Education has always been preparation for life and liberation of the self. "If knowledge continues to increase, the world will need wisdom in the future even more than it does now," said Bertrand Russell. The goal of wisdom is to comprehend the deeper meaning of known facts, and for that, a combination of cognition, self-reflection and openness are necessary. Wisdom is about knowing what to overlook. Wisdom is understood at the experiential level.

We are observing momentous developments in the sphere of science and technology. The question we ask is: Should we leave the destiny of man in the hands of science alone? Man belongs to two different worlds – the world of natural necessity, and the world of moral freedom. A proper balance between the two will be the major challenge for man in the coming years.

We are moving very fast in the material world, and if we want to avoid vortex formation in our lives, we must place moral baffles at appropriate places in the fluid-filled bucket, called life.

Man was once the weakest animal on this earth. He wanted to survive, yet he did not have inherent means for survival. The armoury he developed for his survival included capacity for thought, imagination, ingenuity, and self-awareness. Man is now the most evolved species. With the development of reason and conscience, he emerged from nature; he is now 'fully born', 'fully awake', and 'fully human'. He is the sum of 'inherited' and 'acquired' qualities. He is product of gene and society. He now wants to rewrite life.

In many respects man is still inadequate. Though, man knows, how a complete man should look, man can't engineer a complete man. "When systems are not engineered but instead allowed to evolve – to build themselves – then the resultant whole is greater than the sum of its parts." We, the humans, must take the cognizance of the fact that we are produced by a process that is not engineering. As we know, there is another approach, besides the strict engineering approach, which can produce something of that complexity, and that's the evolutionary approach.

This book is for everyone, as everyone is connected with Science, Engineering and Technology (SET) in some way or the other. SET impacts society and culture. The philosophy of SET accepts "the primacy of the humanities over technologies."

SET has evolved bit by bit, and is still evolving, to meet its own requirements. The SET mindset connects our inner self with the outer nature. It teaches us the benefits of sharing. This book is about connectivity and continuum. This book is about optimum optimism.

Science is a way to understand nature. Engineering aims at modifying the nature. The management of change is central to the practice of engineering. Technology transforms society. The genesis of society is linked to the intellectual and social consequences of science and technology.

Science refines the mind, and vice-versa. Engineering modifies nature, and vice-versa. There is so much to learn from nature. Nature helps to design the modifications. Engineers are, therefore, required to use restraint while modifying the nature.

On a pious occasion, I got the following message from several sources.

"The demon is always within.
The goddess is always within.
The battle too, is always within.
And so is the triumph of one, over the other.
Which one, over which one?
That choice is also, somehow always within you."

For me it was the Governing Principles of Engineering. The other two are: your neural suitcase defines the size of your world, and there is no better engineering than life itself.

May 28, 2020 **Purnendu Ghosh**

Contents

Glossary

A

Additive Manufacturing (AM): The technologies that build 3D objects by adding layer-upon-layer of material. Once a Computer Aided Design (CAD) sketch is produced, the AM equipment reads in data from the CAD file and lays downs or adds successive layers of liquid, powder, sheet material or other, in a layer-upon-layer fashion to fabricate a 3D object.

Adjacent Possible: It is a kind of shadow future, hovering on the edges of the present state of things, a map of all the ways in which the present can reinvent itself. It captures both the limits and the creative potential of change and innovation.

Artificial Intelligence (AI): It is the simulation of human intelligence processes by machines, especially computer systems. The processes include learning, reasoning and self-correction.

Augmented Reality (AR): An interactive experience of a real-world environment where the objects that reside in the real world are enhanced by computer-generated perceptual information, sometimes across multiple sensory modalities, including visual, auditory, haptic, somatosensory and olfactory. AR alters one's ongoing perception of a real-world environment, whereas virtual reality completely replaces the user's real-world environment with a simulated one.

Autonomous Robot: A robot that performs behaviours or tasks with a high degree of autonomy. Autonomous robotics is a subfield of artificial intelligence, robotics, and information engineering.

B

Big Data Analytics: Process of examining large and varied data sets to uncover information — such as hidden patterns, unknown correlations, market trends and customer preferences — that can help organizations make informed business decisions.

Biocompatibility: It refers to the properties of materials being biologically compatible and thus accepted by the biological system so that it does not cause any toxic effects or any elicitation of any response from a living system or tissue.

Biodiversity: It refers to total variability among living organisms (which include genes, species and ecosystems) of a specific region or in a particular habitat.

Bioinformatics: It is the application of information technology to life sciences. It involves the collection, storage and analysis of biological information using computers.

Biological Computing: Also known as Biocomputing or Computational Biology, it refers to the computational solutions to the biological problems through informatics approaches.

Biological destiny: There is a view that says that human behaviour is to some significant degree determined by our biological inheritance. It claims that human evolution has made such characteristics as social hierarchy, gender inequality, competition, and violence inevitable features of every social system. It claims that significant aspects of human behaviour are genetically determined. It says that our fate is in our genes. In other words, genetic research can help to eliminate key social problems. The problems lie not in the structure of society, but in some of the individuals who make up society. Freud wrote that 'Biology is destiny'. Many, however, argue that biology is not destiny. This is a hot topic of debate.

Biologism: The use of biological explanations in the analysis of social situations.

Biomanufacturing: It utilizes biological systems to produce commercially important biomaterials and biomolecules for use in medicines, food and beverage processing, and industrial applications.

Biomarker: A biological characteristic that is objectively measured and evaluated as an indicator of normal biological or pathological processes, or a response to a therapeutic intervention.

Biomaterials: A biomaterial is any natural or synthetic substance that has been engineered to interact with biological systems for a specific purpose - either a therapeutic (treat, augment, repair or replace a tissue function of the body) or a diagnostic one.

Bioprocessing: A technique that uses whole organism or its parts (e.g., bacteria, enzymes, chloroplasts) to obtain desired products (chemical, antibiotic, bioenergy, etc) or biological materials.

Blockchain Technology: Typically refers to the transparent, publicly accessible ledger that allows us to securely transfer the ownership of units of value using public key encryption and proof of work methods.

Brain Mapping: Brain mapping attempts to provide a complete picture of the brain's structure, that is, brain regions with proper functional assignments.

C

Cancerization: The replacement of the normal cell population by a cancer-primed cell population that may show no morphological change is now recognized to underlie the development of many types of cancer, including the common carcinomas of the lung, colon, skin, prostate and bladder.

Cephalization: It is an evolutionary trend in which, over many generations, the mouth, sense organs, and nerve ganglia become concentrated at the front end of an animal, producing a head region. This is associated with movement and bilateral symmetry, such that the animal has a definite head end.

Cognitive Miser: Tendency of people to think and solve problems in simpler and less effortful ways rather than in more sophisticated and more effortful ways, regardless of intelligence.

Cognitive Neuroscience: The study of the biological processes and aspects that underlie cognition, with a specific focus on the neural connections in the brain which are involved in mental processes. It determines how neurological mechanisms are involved in thinking and behaviour.

Confabulation: A neuropsychiatric disorder wherein a patient generates a false memory without the intention of deceit. It is a kind of 'honest lying'. The hypothesis is that the patient generates information as a compensatory mechanism to fill holes in one's memories.

Conscious and Unconscious Learning Processes: Learning processes mainly refer to the interconnections between perception, memory, language, imagery, emotion, and motivation that allows one to mentally build connections between verbal and pictorial information patterns or between new and prior memories and integrate them with relevant knowledge structures in long-term memory. Learning processes and outcomes can be conscious and unconscious. The unconscious processes range from registering information in the sensory memory to mentally forming associations within or between information patterns and activating associative memory networks, including individual expectations, beliefs, and desires. On the contrary, a conscious learning process starts by deliberately paying attention to instructional materials, noticing similarities and differences between words and their particular meanings with the help of relevant prior experience, thereby mentally building coherent connections between them and organizing them into new knowledge structures. Thus, either conscious or unconscious learning is primarily a combination of mental processes, referred to as a knowledge acquisition process, bringing memories into the mind, forming associations, retaining, and using them.

Cybersecurity: It refers to a set of techniques used to protect the integrity of an organization's security architecture and safeguard its data against attack, damage or unauthorized access.

Cyborg: Cybernetic organism is a being with both organic and biomechatronic body parts.

D

Data Mining: It refers to a process of systematic analysis of data for discovering patterns or relationships which gives useful information or knowledge about behaviour of data.

Designed Genomics: It refers to the science of designing and building a minimal genome or specified genome with particular characteristics.

DNA: A self-replicating material present in all living organisms; it is the carrier of genetic information which is responsible for all traits or physical characteristics.

Downstream Processing: It refers to the recovery and purification of biosynthetic products. It is achieved through a series of steps for the concentration, extraction, purification and recovery of useful products from a fermentation reaction.

Drone: An unmanned aerial vehicle; the flight of Drones may operate with various degrees of autonomy: either under remote control by a human operator or autonomously by on-board computers.

E

Enzyme: An enzyme is a protein manufactured by a cell which acts as a catalyst in various biochemical reactions. It is also known as biocatalyst.

Evolutionary Genomics: It involves the study of genome evolution with respect to time. Evolution of a genome is a continuous process by which a genome accepts changes in its structure (sequence) or size over time.

'Fixed' and 'Growth' Mindset: A fixed mindset assumes that our character, intelligence, and creative ability are static givens which we can't change in any meaningful way, and success is the affirmation of that inherent intelligence. A growth mindset thrives on challenge and sees failure not as evidence of unintelligence but as a heartening springboard for growth and for stretching our existing abilities. These two mindsets decide our relationship with success and failure in both professional and personal contexts.

F

Functional Genomics: It is a branch of genomics that uses genomic data to study gene and protein expression and function on a global scale (genome-wide or system-wide), focusing on gene transcription, translation and protein-protein interactions, and often involving high-throughput methods.

G

Gene expression: Gene, a unit of inheritance, comprises of a stretch of DNA that codes for a protein. Central Dogma explains the flow of genetic information within biological system i.e., DNA encodes RNA (transcription) and RNA encodes Protein (translation). This process of gene expression creates the building blocks of life from genetic information.

Gene Therapy: Gene therapy involves the insertion of genes (exogenous DNA) into an individual's cells and tissues to treat a disease, and hereditary diseases in which a defective mutant allele is replaced with a functional one. In most cases of gene therapy, a "normal" gene is inserted into the genome to replace an "abnormal," disease-causing gene.

Genetic Engineering: It is manipulation of an organism's genome using various tools of biotechnology to produce improved or novel organisms.

Genome Editing: It refers to editing (add, remove or alter) the nucleotides of the genome with engineered nucleases in cultured cells or living organisms. By editing the genome the characteristics of a cell or an organism can be changed. In the past decade, several types of engineered nucleases have been developed, including zinc finger nucleases (ZFNs), transcription activator-like effector nucleases (TALENs), and the recent clustered regularly-interspaced short palindromic repeat (CRISPR) systems. Genome editing holds great promise in potential clinical applications, such as gene therapy.

Genome: It refers to the total genetic material present in a single cell of an organism. Each genome contains the complete set of genetic information required to build and maintain that organism.

Genomics: The study and analysis of all genes and their inter relationships in an organism, so as to identify their combined influence on its growth and development.

Geography of Thought: This book by Richard Nisbett tells how Asians and Westerners think differently. The book argues that cultural differences affect people's thought processes more significantly than believed. The book assumes that human behaviour is not "hard-wired" but a function of culture.

Gut Metagenome: It is the aggregate of all the genomes of gut microbiota.

H

Homologous Recombination: It is a type of genetic recombination in which nucleotide sequences are exchanged between two similar or identical molecules of DNA. It is most widely used by cells to accurately repair harmful breaks that occur on both strands of DNA, known as double-strand breaks.

Human Brain Project: It is a global initiative for reconstructing and simulating the human brain to accelerate the fields of neuroscience, computing and brain-related medicine. The project is focusing on Neuroinformatics, Brain Simulation, High Performance Computing, and Neurorobiotics.

Human Brainome Project: To develop new tools for exploring neural circuitry.

Human Cell Atlas: The agenda of human cell atlas is to create comprehensive reference maps of all human cells as a basis for both understanding human health and diagnosing, monitoring, and treating disease. A complete Human Cell Atlas intends to give a unique ID card for each cell type, a three-dimensional map of how cell types work together to form tissues, knowledge of how all body systems are connected, and insights into how changes in the map underlie health and disease. It would allow us to identify which genes associated with disease are active in our bodies and where, and analyse the regulatory mechanisms that govern the production of different cell types.

Human Enhancement: The application of technology to overcome physical or mental limitations of the body, resulting in the temporary or permanent augmentation of a person's abilities and features.

Human Genome Project: An international scientific project with the goal of determining the sequence of nucleotide base pairs that make up human DNA, and of identifying and mapping all of the genes of the human genome from both a physical and a functional standpoint.

I

Internet of Things: The term refers to scenarios where network connectivity and computing capability extends to objects, sensors and everyday items not normally considered computers, allowing these devices to generate, exchange and consume data with minimal human intervention. Before its full potential is realized, it has to overcome several challenges, like, surveillance concerns and privacy fears.

W

Mental Simulation: It is our mind's ability to imagine and simulating the probable result before taking a specific action. It relies on our memory, perception and experience. Its conscious harness can be extremely powerful. It dramatically enhances our ability to solve novel problems. Research suggests that the tendency for mental simulation is associated with enhanced meaning.

Metabolic Diseases: Diseases caused due to any disruption in metabolic processes.

Metabolic Engineering: The use of genetic engineering to modify the metabolism of an organism. It can involve the optimization of existing biochemical pathways or the introduction of pathway components, most commonly in bacteria, yeast or plants, with the goal of high-yield production of specific metabolites for medicine or biotechnology.

Metabolomics: It refers to the systematic identification and quantification of the small molecule metabolic products (the metabolome) of a biological system (cell, tissue, organ, biological fluid, or organism) at a specific point in time. Mass spectrometry and NMR spectroscopy are the techniques most often used for metabolome profiling.

Metagenomics: The study of the metagenome—the collective genome of microorganisms from an environmental sample—to provide information on the microbial diversity and ecology of a specific environment.

Microbiome: The totality of microorganisms and their collective genetic material (or genome) present in particular environmental niche, for example human microbiome (means in entire human body) .

Microbiota: A microbiota is an ecological community of commensal, symbiotic and pathogenic microorganisms found in and on all multicellular organisms studied to date from plants to animals.

Minimum Genome Project: The project aims to search for minimal base gene set required for life. It identifies the "Minimal Genome Organism".

MOOCS: Massive Open Online Courses (MOOCs) are free online courses available for anyone to enrol. MOOCs provide an affordable and flexible way to learn new skills, and deliver quality educational experiences at scale.

Mutation: A change in a genetic sequence; includes changes as small as the substitution of a single DNA building block, or nucleotide base, with another nucleotide base. Along with substitutions, mutations can also be caused by insertions, deletions, or duplications of DNA sequences.

N

Nanobiology: A branch of biology dealing with nanoscale biological interactions. It is a field of study, that merges biological research with nanotechnologies, such as nanodevices, nanoparticles, or unique nanoscale phenomena.

Nanomedicine: A branch of medicine that applies the knowledge and tools of nanotechnology for the prevention and treatment of diseases. It involves the use of nanoscale materials, such as biocompatible nanoparticles and nanorobots, for diagnosis, delivery, sensing or actuation purposes in a living organism.

Neural Implants: Surgically implanting the neural devices in the brain for research purpose, or for the treatment of neurological disease and brain injuries.

Neurobiotechnology: It is the scientific study of the nervous system using various molecular tools of biotechnology.

Neurogenetics: A branch of genetics dealing with the study of genetic factors that contributes to the development of nervous systems and genetic disorders responsible for neurological disorders.

Neuroimaging: It deals with the *in vivo* depiction of anatomy and function of the central nervous system in health and disease. Neuroimaging is an essential adjunct to clinical and cognitive assessments in the evaluation of cognitive and behavioural disorders.

Neurons: Considered the basic units of the nervous system, Neurons, using electrical and chemical signals, help coordinate necessary functions of life. Sensing what is going around us, neurons decide how we should act, alter the state of internal organs, and allows us to remember what is going on. There are around 90 billion neurons in the brain; each neuron is connected to another 1,000 neurons.

Neuroplasticity: The brain's ability to form new neural connections. It allows the neurons in the brain to compensate for injury and diseases, and to adjust their activities in response to new situations or changes in their environment.

O

Oncogenes: A gene that has the potential to cause cancer.

P

Personalized Therapy - Therapy based on the context of a patient's genetic content or other molecular or cellular analysis.

Phage Therapy: The therapeutic use of lytic bacteriophages to treat pathogenic bacterial infections.

Phage: A short term for Bacteriophage, which is a virus that infects and replicates within a bacterium.

Pharmacogenomics: Pharmacogenomics is the study of how genes affect a person's response to drugs. This relatively new field combines pharmacology (the science of drugs) and genomics (the study of genes and their functions) to develop effective, safe medications and doses that will be tailored to a person's genetic makeup.

Protein Design: A technique by which proteins with enhanced or novel functional properties are created. It is the rational design of new protein molecules to fold to a target protein structure with desired functionality.

Protein Expression Profiling: Identifying the proteins expressed in a particular tissue, under a specified set of conditions and at a particular time, usually compared to expression in reference samples. Through proteomics study, we try to measure the thousands of expressed proteins in a single experiment.

Proteomics: It is the study of large-scale analysis of Proteome (total expressed proteins in a cell under given condition) with respect to diseases and environmental conditions. It is mainly used for comparing protein structure (structural proteomics), protein expression (expression proteomics), and for characterizing protein-protein interactions (interaction proteomics).

Q

Quantum Cryptography: The science of exploiting quantum mechanical properties to perform cryptographic tasks. Cryptography is about constructing and analysing protocols that prevent third parties or the public from reading private messages; various aspects in information security such as data confidentiality, data integrity, authentication, and non-repudiation are central to modern cryptography.

R

Regenerative Medicines: A branch of translational research in tissue engineering and molecular biology which deals with the "process of replacing, engineering or regenerating human cells, tissues or organs to restore or establish normal function".

Reverse Engineering the Brain: It is an intersection of artificial intelligence (AI) and neuroscience to study how the brain learns and constructs logical rules, and how its performance of those tasks compares with that of the artificial neural networks used in AI.

S

Sadi Carnot: A French military scientist and physicist (1796 – 1832) put forward the first theory of the maximum efficiency of heat engines. He is known as the father of thermodynamics. His work helped in the formalization of the second law of thermodynamics and define the concept of entropy. Carnot's only book published in 1824 - Reflections on the Motive Power of Fire - received very little attention from his contemporaries. Clausius and Kelvin elaborated upon his work. His work is now a classic.

Scale Down: It is a procedure mainly used for troubleshooting or development of processes used in ongoing production whereby we try to reduce or minimize the cost of a process. In scale-down, we try to use similar parameters (like, fermenter geometry) that we use at larger scale.

Scale Up: It is a procedure of moving from smaller scale production to larger scale production.

Self-organized Criticality: It is a phenomenon observed in certain complex systems of multiple interacting components, e.g., neural networks, forest fires, and power grids, that produce power-law distributed avalanche sizes.

Single Nucleotide Polymorphisms (SNPs): It is a DNA sequence variation occurring when a single nucleotide (A, T, G or C) in the genome (or other shared sequence) differs between members of a species or paired chromosomes in an individual.

Singularity: A hypothetical future point in time when technological growth becomes uncontrollable and irreversible, resulting in unfathomable changes to human civilization. Ray Kurzweil believes that by the year 2045 we will experience the greatest technological singularity in the history of mankind. It will change the pillars of society completely and the way we view ourselves as human beings. In a way it suggests merger of humans and robots.

Space-time Yields: Space-time is the time necessary to process one reactor volume of fluid, at a given set of conditions. Space-Time Yield is the amount of product synthesized per reactor volume per unit time.

Stem Cell: It is an undifferentiated cell of a multicellular organism which is capable of giving rise to indefinitely more cells of the same type, and from which certain other kinds of cell arise by differentiation.

Synapse: The word 'synapse' means 'conjunction'. Being an essential neuronal function, Synapse links one neuron to another. It is a structure in the nervous system that permits a neuron to pass an electrical or chemical signal to another neuron. When a nerve impulse reaches the synapse at the end of a neuron, it triggers the neuron to release a chemical neurotransmitter. The neurotransmitter drifts across the gap between the two neurons, and on reaching the other side, it fits into a tailor-made receptor on the surface of the target neuron, like a key in a lock. This docking process converts the chemical signal back into an electrical nerve impulse.

Synthetic Biology: Synthetic biology seeks to model and construct biological components, functions and organisms that do not exist in nature, or to redesign existing biological systems to perform new functions.

Synthetic Genomics: It encompasses technologies for the generation of chemically-synthesized whole genomes or larger parts of genomes, allowing to simultaneously engineering a myriad of changes to the genetic material of organisms, for example, artificial gene.

Systems Biology: It is a holistic approach to decipher the complexity of biological systems that starts from the understanding that the networks that form the whole of living organisms are more than the sum of their parts. It integrates many scientific disciplines – biology, computer science, engineering, bioinformatics, physics and others – to predict how these systems change over time and under varying conditions, and to develop solutions to the world's most pressing health and environmental issues.

Systeomics: It is defined as the integration of genomics, proteomics, and metabolomics.

T

Technium: Kevin Kelly thinks that the relationship between humans and technology have become so complex and interwoven with our lives that humans have less and less sway over how mechanical systems evolve. Kelly introduces the concept of the 'technium' to embody the vast techno-social system that includes all the machines, processes, society, culture and philosophies associated with technologies. The sheer complexity of interactions between the various layers and loops of the technium gives it a degree of autonomy. As it evolves, it develops its own dynamics.

The Cloud: A global network of servers, each with a unique function, and meant to operate as a single ecosystem. These servers are designed to either store and manage data, run applications or deliver content or a service such as streaming videos, web mail, office productivity software or social media. Businesses use four different methods to deploy cloud resources. There is a public cloud which shares resources and offers services to the public over the Internet, a private cloud which is not shared and offers services over a private internal network typically hosted on-premises, a hybrid cloud which shares services between public and private clouds depending on their purpose and a community cloud which shares resources only between organisations, such as with government institutions.

Thought experiment: "A thought experiment is a device with which one performs an intentional, structured process of intellectual deliberation in order to speculate, within a specifiable problem domain, about potential consequents (or antecedents) for a designated antecedent (or consequent)". One of the objectives of thought experiments is to gain new information by rearranging or reorganizing already known empirical data in a new way and drawing new inferences from them or by looking at these data from a different and unusual perspective.

Tissue Engineering: Tissue engineering is a set of methods that can replace or repair damaged or diseased tissues with natural, synthetic, or semisynthetic tissue mimics. These mimics can either be fully functional, or will grow into the required functionality.

Transcriptomics: Study of the transcriptome - the complete set of RNA transcripts that are produced by the genome, under specific circumstances or in a specific cell—using high-throughput methods, such as microarray analysis.

T-shaped: A metaphor used in job recruitment to describe the abilities of persons in the workforce. The vertical bar on the letter T represents the depth of related skills and expertise in a single field, whereas the horizontal bar is the ability to collaborate across disciplines with experts in other areas and to apply knowledge in areas of expertise other than one's own.

Two Culture: The Two Cultures, C. P. Snow's hypothesis, suggests that "the intellectual life of the whole of western society" was split into two cultures – the sciences and the humanities – which was a major hindrance to solving the world's problems.

W

'Will' and 'Grace' Hypothesis: According to the "Will" hypothesis, honesty results from the active resistance of temptation. According to the "Grace" hypothesis, honesty results from the absence of temptation.

1

Mindset Defines the Contours of Life

Neural suitcase determines the size of our world.

We think. We imagine. We generate ideas. We make sense of things. We take decisions. We observe patterns, and then try to understand their significance. Our neural suitcase does all this. It tells us the tales of many minds (1). The tales are beautiful, vulnerable, quiet, chaotic, real, fictional, wise, foolish, friendly, hateful, meaningful, and blind. Our neural suitcase builds castles in the air. It determines the size of our world.

The Brain

It is our brain that helps us to prioritise and set our goals. Brain receives signals in various forms. Our emotions influence these signals. Due to the clashes in emotions, the likelihood of clashes in our drives always exists. Our brain sorts out the priority, based on our past experiences, skill sets, and cultural leanings. Our brain sees what matters to us the most, and then we try to reach there. It also takes us away from irrelevant reaches.

Brain is a physical entity. Brain interprets our senses, initiates our body movements, and controls our behaviour. The human brain has evolved over time; human brain size tripled over the course of human evolution. As the structure of the brain changes, it changes our behaviour. The building blocks that made the brain existed for billions of years in the ocean. Sponges, one of the earliest groups of animals, do not have nervous systems. Jellyfish perhaps is the first group of animals to evolve genuine neurons, but these neurons were arranged in a diffuse net. There was no central processor. The next step in the evolutionary history of the brain was a process known as 'cephalization' in which neurons cluster at one end of an animal, eventually becoming a brain linked to important sensory organs like eyes. Brains then began to specialize in different functions. The enlargement of the brain resulted in the emergence of more sophisticated intelligence. "Even before life left the water, animals had evolved brains with much of the same basic neural architecture that we would eventually inherit," writes Ferris Jabr (2).

There are so many levels to understand the brain – genes, proteins, cells, circuits, animal behaviour. Neuroimaging methods, in combination with genomic data, are needed to develop an extensive map of how the brain works. These maps tell us more accurately how a normal or a disrupted brain works.

The human brain is an intricate network of 100 billion neurons, and 100 million synapses. The wiring diagram of the brain helps us to understand how the brain functions, and also sheds light on disorders that are presumed to originate from faulty wiring. The wiring diagram of the brain is not easy to draw. Scientists have been able to draw the invaluable wiring diagram of *C. elegans*, a microscopic worm containing a mere 302 neurons. It took the neuroscientists more than a decade to draw the diagram of this tiny worm. In spite of the difficulties envisaged, scientists are hopeful of completing the human brain's wiring diagram in a foreseeable future. The technological advance in neuroscience is one of the reasons of their optimism.

Synapses are essential to neuronal function. There is substantial rearrangement and pruning of synapses during brain development and growth. Experience influences the rewiring of synapses during brain maturation. This rewiring is not just limited to the young brain. Synapses are added, as well as lost; the peak is between six and eight months postnatal. The initial synapses formation is independent of stimulation. But if the synapses are not used, the brain eliminates them. Conversely, more often a connection is used, stronger it becomes in a physical sense, with more dendritic spines connecting to one another resulting in a stronger net connection over time.

The capacity of the brain to change its connections is dependent on genetic regulation. This understanding raises a number of interesting questions. For example, how the brain can be so plastic, and yet can retain memories over time. To answer such questions one would have to address the underlying genetics. Fortunately new gene-profiling tools now available can figure how neurons are manifest in the body.

Understanding how genes and experience come together to impact the brain could significantly alter how we think. The advent of high-throughput gene profiling and the growing sophistication of our ability to manipulate genes in animal models have given us an opportunity to explore the role that genes play in both creating and modulating our neural structures. The new imaging techniques and technologies have made us better equipped to characterise neural systems and their response to the world around us.

Genes don't tell the whole story. Environment is needed to complete the story. Our experience can change the genetic expression and thus the story. Neuroscience says that actions can occur even when the will to act is missing.

Some people can't override or control what subconscious parts of their brains decide to do. Neurobiologists say that no part in the brain is 'independent' and therefore 'free'. They say, even if free-will exists it can at best be a small factor riding on top of vast neural networks shaped by genes and environment.

The brain of humankind is not completely formed at the time of birth. It changes as it goes along, probably to adapt to the changing environment. The average new-born human brain weighs less than 400 g. A typical human adult brain weighs about 1400 g. Much of the brain weight increase is during the first three years after birth. After reaching an age of about 20 years, the brain weight starts to decline, slowly but steadily.

Our development in the early years is both robust and vulnerable. Foundational age (from the time we are born to the age 5) is the period when a child's linguistic and cognitive abilities are developed. During this period, they exhibit dramatic progress in their emotional, social, regulatory, and moral capacities. This is the period when one can gauge a child's strengths and weaknesses. This is the period when a child's literacy and numerical skills are developed. This is the period when emotional impairments affect the child quite significantly.

Mahabharata tells the story of Abhimanyu, the son of Arjun and Subhadra. Abhimanyu learned the way to enter the 'Chakravyuha' when he was in his mother's womb. Neuroscience research suggests that the mystery of prenatal consciousness is not a fantasy. It says that foetus knows and learns much more than we presume they know. Neuroscientists believe that consciousness exists from the very first moment of conception. This emerging view is giving new insights into the meaning and responsibilities of parenthood.

Parents are 'active ingredients' of environmental influence during the early-childhood period. The consequences of sour relationship between the parents on a child's development can be severe and long- lasting. Family violence and violent neighbourhood can have demoralizing fallout on the child. A loving family environment is very important for a child's development. It helps boost a child's vocabulary. It makes a child more emotionally secure.

Teenage Brain

Teenage is not simply the continuation of the childhood. It is one of the most interesting phases of brain development. It is the phase that is developing, yet not fully ready to face the challenges of the complex world. During this phase teenagers develop their identity, and discover things for themselves. It is the time for them to take risks. It is the time when all kinds of accidents happen. It is the time when peer influence hits its peak. It is the time of heightened self-consciousness. It is the time for the onset of anxiety, depression, addiction, and eating disorder. In short, it is the most paradoxical time of one's lifecycle.

It is possible that at different ages we use different brain circuitry and cognitive strategy to perform a given task. It is likely, as adults our decisions to deal with social situations are based more on the prevailing social scenario and norms. Adolescents rely more on their own experiences and gut feeling. The grey matter increases during childhood; it peaks during 11 to 12 years of age, then declines during the period of adolescence right into the 20s or even the 30s. The young brains undergo a massive reorganisation during 12 to 25 years of age. The brain plasticity also makes adolescent brains more vulnerable to external stressors. Young, no doubt have very sharp brains, but they are not quite sure what to do with it. In calm situations, teenagers can rationalise almost as well as adults, but stress can hijack their decision-making.

Cognitive scientists say that a "hurriedly jam-packing a brain is akin to speed-packing a cheap suitcase." Quick learners can hold the load only for a while. A carefully and gradually packed neural suitcase holds its contents much better, and prepares it for better recall. We have known and experienced that harder it is to remember something more difficult it is to forget later.

Our brain is a mix of empathy and system. According to the empathising-systemising (E-S) theory, in some individuals empathising (E) is stronger than systemising (S), whereas in some S is stronger than E. In a "balanced brain" both E and S are equally strong. Some brains may have impaired empathising alongside normal or even talented systemising. Some brains may have impairments in systemising, alongside normal or even talented empathising. Impaired empathising is difficult to handle. Such people are generally misunderstood and mistreated, because they lack empathy. But, if nurtured properly, systemising can be a valuable tool, and can even result in a refreshingly original way of thinking and seeing the world.

Our brain not only thinks, it constantly keeps a tab on how we think. It is always eager to check if the progress of our pursuit is satisfactory or not. It takes the path to reach the goal, and at the same time keeps handy another follow-up path, lest the earlier path fails to take us to the destination. This follow-up thought is useful in many situations. Take the case of insomnia or anxiety, more one wants to avoid anxiety, deeper one gets entangled in it. More one tries to sleep, further one goes away from sleep.

The insomniac's brain keeps on checking if it has fallen asleep or not. The fear of sleeplessness results in a hyper intention to fall asleep. This hyper intention incapacitates one's ability to sleep. It is a kind of fear. To overcome this fear, one should not try to sleep, rather try to do just the opposite, try to stay awake.

In other words, "the hyper-intention to fall asleep, arising from the anticipatory anxiety of not being able to do so, must be replaced by the paradoxical intention not to fall asleep, which will soon be followed by sleep."

Brain is like a toolbox with random tools. You take out one particular tool to solve a specific problem. The problem is how so many tools interact to solve an interrelated problem.

The Mind

Mind is a virtual reality. Mind has awareness and consciousness. Because of the mind we can make sense of our world in a meaningful way. We have the ability to remember. It is because of this ability we can preserve, retain, and subsequently recall knowledge, information, and experience. Mind does things that enable consciousness, perception, thinking, feeling, judgment, and memory.

The most sought after minds, writes Howard Gardner, are the disciplined mind, the synthesising mind, the creative mind, the respectful mind, and the ethical mind. For a disciplined mind, it is not enough to accumulate factual knowledge. A disciplined mind is equipped in specific ways of thinking. It recognises the difference between subject matter and discipline. A synthesising mind can bind part information into a coherent whole. A creative mind goes beyond normal knowledge, poses new questions, and offers new solutions. Too disciplined minds are often in conflict with synthesising and creative minds. A synthesising mind's approach is to seek order, equilibrium, and closure. The creative mind is motivated by uncertainty, surprise, and continual challenge. The respectful mind respects, acknowledges, and views sympathetically and constructively the differences between people.

Mindset plays an important role in shaping our behavioural patterns and decision-taking. We take different decisions for different situations. Some decisions rely on reasons and some on emotions. The 'art of self-overhearing' is one way to take a decision. This 'art' requires our willingness to engage in introspection when one is confronted with an uncertain situation. Such situations often don't give us enough time for 'doing maths'. Decisions in such situations, therefore, depend upon our emotions, instincts and mental short-cuts.

For a decision, both intuition and reason are necessary. Reason enables us to go beyond mere perception, habit, and instinct. Intuition is an inference that is validated by the thinker's belief systems. Some believe that reason-based decisions are better. We also know that many reason-based decisions often go wrong. It is because we systematically look for arguments to justify our own beliefs or actions. We tend to rebut genuine information put forward by others. We forget that, in order to pursue truth, the inputs of others are equally essential. We can't pursue truth by becoming both the judge and the advocate.

A study on the psychology of reasoning suggests that we reason rather poorly; our reason-based decision making are too subjective. Our decisions fail due to our biases. Decisions of even skilled arguers often fail. One way to get over 'biased reasoning' is to become more objective. We must recognize the fact that all arguments are not debates (where the purpose is to win). Some arguments are based on truth, not necessarily to win or lose. Experience, in many situations, is helpful in building patterns, based upon which we can quickly size up the situation and take decisions. Experience teaches us what to ignore, what to watch, and what to expect next.

Intuition is sudden flash of insight. Intuitive decisions are based on understanding, rather than knowledge. One wouldn't know, in many instances, from where the insight emerged. It is our 'gut feeling'. It could be due to 'condensed reasoning'. It could be the result of 'unconscious associative processes'. Emotions are integral part of intuitions. Intuitions often outperform rational analyses.

Our sub-optimal decisions are often a result of our bias. Often, skilled arguers are not after the truth, but after arguments supporting their views. The result is that bias creeps in the reasoning, and it becomes so motivated that it distorts evaluations and attitudes. On the other hand, intuitive decisions, even when based on pure guesswork, are often right. Intuitive decisions are rarely based on mere intuitions. Intuitions are always laced with some kind of mental simulation; our mental simulations do the analysis. Intuition helps us to better our pattern-recognition capabilities and alerts us to possible dangers.

The essence of decision-making is mental simulation. Mental simulation means, "you are trying to predict the outcome before you take an action by using analogies of your previous experiences, or by observing and remembering the outcomes of other people's behaviours." We do mental simulations for many different actions before we actually make a choice. Rewards are often the major parameter that affects our decisions. We take decisions in the hope of better rewards; reward could be immediate or delayed.

It is important to understand how different parts of the brain do different computations in a coordinated way. We need to understand how cognition and emotion work in the brain to produce a decision, and how they become dysfunctional. We need to understand why different individuals make different decisions, and why they make different choices when they face the same situation. Are the differences due to genetic differences, or due to the differences in experiences and learning environments, or both? It is said that decisions taken in a happy state of mind are happy decisions. Trustworthy decisions emanate from trustworthy minds.

Matters related to trust (including trusting the self) occupy a huge amount of our mental space. David DeSteno (3) makes an important observation, "Although it's true that cooperation and vulnerability require two parties, no one ever said that the two parties had to be different people. To the contrary, the parties can be the same person at different times." Being realistic, honest, and forgiving with yourself and others are helpful to deal with the challenges of tough times. Trust has inherent risks, but it is a kind of risk worth taking.

There are trusts that are 'calculation-based'; these calculate the value of creating and sustaining trust relative to the costs of sustaining or severing the trust. The 'identification-based trust', on the other hand, values other person's identity as an individual. For obvious reasons, identification-based trusts are more trustworthy and sustainable. In the 'benevolence-based trust', an individual does not intentionally harm another when is given an opportunity to do so. The other kind is 'competence-based' trust. In this, an individual believes that another person is knowledgeable about a given subject area, and, therefore, can be trusted. Competence and reliability are the two traits that determine and establish trustworthiness. Under vulnerable circumstances we tend to depend more upon trusts. We shift from one mode of trust to the other, depending upon the circumstances.

One of the essential ingredients of knowledge-based systems is knowledge-sharing. Knowledge-sharing is not possible unless trust is embedded in the system. Trust evokes a feeling of confidence. A trusted person enjoys many advantages; even the vulnerability of a trusted person is accepted. A person can be trusted in some contexts, but not necessarily in all the contexts. Distrust, on the other hand, evokes a feeling of doubt and fear. The reasons of distrust could be purely imaginary, resulting out of misplaced feelings. Such feelings unnecessarily create vulnerability.

A little distrust is often helpful. It can avert herd mentality. Relationship experts say that healthy amount of distrust can be protective against the risk of exploitation. Does trust need periodic validity checks? Yes, it does, if the trust is event-based, as the continuance of event-based trusts generally depend on future events. If the trust is process-based, it is more likely to become permanent.

The areas of the brain that register physical pain are active when someone feels socially rejected. Social pain can trigger the same sort of distress as stomach ache, or broken bone triggers. When brain knows that we are with someone trustworthy, it allocates precious resources more wisely. Instead of using it for coping with stressors or menace, it uses it in learning new things or fine-tuning the process of healing.

Our biases and prejudices are subjective. We rely more on a certain piece of information, as we think other alternatives might not work that effectively. Because of distorted perceptions and wishful thinking, we see things more positively than they really are. We even distort our memory to suit our perception.

We all have 'bias blind spots'. We take biased decisions, based upon the irrational decisions we have taken in the past. We accept things, not necessarily based on merit, but merely because of our familiarity with the thing. We take hasty decisions to escape the feeling of doubt and uncertainty. We overestimate the degree to which others should agree with us. We underestimate others' ability to understand us. We overestimate our own ability to know others. What was said matters less to us than who said it.

Our brains are attuned to negative biases. Majority of our decision making biases favour conflict, rather than concession. Hawks see only hostility in their adversaries; Doves often point to subtle openings for dialogue. In fact, a bias in favour of hawkish beliefs and preferences is built into the fabric of the human mind, says Daniel Kahneman (4). Biases don't die; they take new forms. We find it almost impossible to ignore the "subjective first person view of things." All biased decisions, however, are not necessarily unfair and bad.

Then there is conformity to peer pressure. We accept the arguments of those whom we like, and reject the arguments of those whom we dislike. We attribute our success to our abilities and talents, and our failures to bad luck, external factors and destiny. We use one yardstick to judge success/ failures of the self, and another yardstick for others. We underestimate the influence of self-interest on our own judgments and decisions, but overestimate its influence on others. We underestimate future uncertainties. We cross-check the bad news, but readily accept the good news.

Decision making is a subjective perception and experience. Some of us like familiarity and certainty. Some of us like uncertainty and novelty. People in the first group are less receptive to new ideas and are biased towards predictability and clarity. People belonging to the second group love to face new situations and are biased towards such people/issues.

Two microscopes can be used for two different kinds of imaging, thus giving a powerful combination of high specificity and detailed structural information. Taking a cue, does it mean that two minds are better than one? Behavioural neuroscientists say that two people can produce worse judgments, not because together they are not capable of making a good decision, but because of the confidence that they can; togetherness is often the reason of overconfidence. Psychologists call this the 'cost of collaboration'.

"The intuitive mind is a sacred gift, and the rational mind is a faithful servant. We have created a society that honours the servant, and has forgotten the gift," wrote Albert Einstein years ago. Pure rational decision-making is difficult, as our decisions are always coloured with our emotions. Neuroscientists say, the rational mind vacillates endlessly over the possible rational reasons. Both 'cold reasons' and 'hot emotions' are needed to take 'right decisions'.

We possess finite mental energy for exerting our willpower. Willpower is like a muscle that gets fatigued when overused. It gets exhausted when it is repeatedly used in a short span of time. A fatigued decision maker often tries to take the path of least resistance. Often paths of least resistance don't lead us to the desired destinations. The big problem is, it is not easy to know when we are decision-fatigued, and when our willpower is low. We are always eager to take decisions.

When decision fatigue sets in, either we do nothing, or we take short cuts, or we try to maintain status quo. Doing nothing eases mental strain for the time being, but can create problems, may not be immediately, but in the long run. When willpower is depleted, one tends to act as a 'cognitive miser'. If you are shopping, it makes you vulnerable and easy target of the sellers and the marketers. "When you shop till you drop, your willpower drops, too."

Decision fatigue and willpower are special problems for poor people. The poor have less buying power, and, therefore, becomes satisfied easily. Choices generally don't matter much to them. In one study, researchers found that poor people do more 'impulse purchases'. They do this in spite of their lesser spending capacity. The impulse purchase reflects lesser hold on their self-control. "Part of the resistance against making decisions comes from our fear of giving up options." When our willpower becomes weak, our frustrations become more irritating. We develop a "propensity to experience everything more intently."

Some of us like familiarity and certainty. Some of us like uncertainty and novelty. People in the first group are less receptive to new ideas, and are biased towards predictability and clarity. People belonging to the second group love to face new situations and are positively biased towards such people/issues.

The External Mind

We are constantly confronted with the explosion of ideas, but do not possess enough 'brains' to deal with them. The explosion of ideas generates cognitive stress. The management of cognitive stress due to information overload requires enormous self-discipline. It is a myth that constant exposure to new information makes us more creative. Information overload, in fact, has been found to do the opposite. We needs to filter out the unimportant. We need to know what to address and what to delegate. One way to deal with information overload is to keep 'alone time', the time during which one does not want to be connected with others, but wants to get engaged with the self. This is 'focused working session'.

Internet, it is said, is shifting our cognitive functions; from searching for information inside the mind to searching outside the mind. In the process Internet has become our external memory storage system. Some of its consequences include fragmented thinking and shorter attention spans, reduction in reflection, introspection, and in-depth thought. The bigger fear, observers feel, is the extinction of experience with the natural world.

Ernst Poppel beautifully describes this feeling of getting lost in the new horizon. "It is like swimming in an ocean with no visible horizon. Sometimes suddenly an island surfaces unexpectedly indicating a direction, but before I reach this island, it has disappeared again. This feeling to be at a loss has become much stronger with the internet."

So, don't swim in the ocean of information, unless there is direction, and you have an idea of the island. If the internet has made us 'lazy, stupid, lonely, or crazy', it has also made us 'smarter than we've ever been before'. It is fast becoming our 'external hard drive'. Now with the 'extended mind', it is not important to carry such information in mind that are readily available at the external mind.

Our brains are busier than ever before. But our brains are "not wired to multitask well… When people think they're multitasking, they're actually just switching from one task to another very rapidly. And every time they do, there's a cognitive cost in doing so," writes Daniel Levitin (5).

Multitasking, some call 'infomania'; even an unread email can effectively reduce your IQ by 10 points. According to a neuroscientist, "learning information while multitasking causes the new information to go to the wrong part of the brain. If students study and watch TV at the same time, for example, the information from their schoolwork goes into the striatum, a region specialised for storing new procedures and skills, not facts and ideas. Without the distraction of TV, the information goes into the hippocampus, where it is organised and categorised in a variety of ways, making it easier to retrieve."

'Infovores' eat information. When they eat more information than they can digest, they develop 'infobesity'. The major causes of obesity are poor eating habits, both quantitative and qualitative, and reduced physical activity. What matters is not only what one eats, but also how one eats it. Similar logic applies for infobesity. "If we're turning into informavores, it's probably because we want to," observes Nicholas Carr (6). It is important to know what to remember, what not to remember, which idea is stronger, which idea is weaker, which idea would survive, which idea would drown. As one CEO advises, "You have to guard against the danger of over-eating at an interesting intellectual buffet." If one eats smaller portions of information and relishes it, the chances of his developing infobesity are much lesser.

Our mind wanders, because consciousness changes its reference points every moment. It is, therefore, natural for a wandering mind to experience spontaneous, unfocused, and unconstrained thoughts. In those 'unfocussed and unconstrained' moments, it is likely for one to find what one is looking for. In mundane moments our mind generates spontaneous internal thoughts. Some, thus say, building castles in the air is not necessarily a waste of time. Some say that focus has distraction built into it. Distraction often helps in finding new ways, as we love variety, surprise and adventure of the unknown. A wandering mind is not necessarily an unhappy mind. A wandering mind can protect us from immediate perils, and can keep us on the course to reach our long-term goals. It is said that a wandering mind is good for creative activities.

Neuroscience says, action can occur even when the will to act is missing. It is not always possible to override or control what subconscious parts of our brain has decided to do. No part of the brain is 'independent' and therefore 'free'. It means, it is not always necessary to know the purpose or destination before embarking on a journey. Every effect may have a cause, but it is not necessary to know what that cause is. In other words, it is possible to solve new problems independently of previously acquired knowledge.

Our relationships shape our brain, which in turn shape our relationships. Loving relationships alter the brain most significantly. A loving touch can make a big difference. A happy woman in a committed relationship, when given an electric shock, registered less anxiety and pain when holding her partners hand. A lover's touch has been shown to subdue blood pressure, ease responses to stress, and soften physical pain. In not-so -loving relationships, this effect was not observed.

Who Decides Our Moral Decisions?

"To be a moral being is to be capable of being motivated to do what you ought to do because you believe you ought to do it."

Does a moral organ exist in our body? We are not so sure. Some say, our reading histories, novels, philosophical treatises and ethnographies have helped us to update our moral software. May be someday, some say, biological hardware will be used for updating our moral software.

Morality is greatly influenced by our local culture and learning. Honest minds don't intentionally misrepresent facts, intentions, or opinions. Honest people don't get tempted to behave dishonestly. We are generally honest, in spite of the fact that we have resorted to some sort of dishonesty some time or the other. Often we have behaved dishonestly even when our intentions were not so.

The 'Grace Hypothesis' says that the decisions of honest people don't change when they get an opportunity to increase their reward by being dishonest. The `Will Hypothesis', on the other hand, says that honest behaviour can also result from the intentional resistance of perceived temptation to behave dishonestly.

An intelligent and ethical mind is the most desirable combination. On the other hand, combination of an intelligent and an unethical mind is the most disastrous combination. Often intelligence and talent don't go hand-in-hand. We often become so convinced about the rightness of our own ideas that we start believing that we only have the right to be right. We tend to push threatening information away from us, and pull friendly information close to us. We dislike arguments that try to change our opinion. We like arguments that help us to hold our views.

Honesty, often, is not the best policy, as truth hurts. In a strongly competitive environment ambition and vanity sometimes completely outweigh our ethics and sense of fairness. Honesty can be shifty. We can be truthful even when what we say is not actually true. We can deceive and mislead others without telling a lie. Keeping quiet when one supposedly should speak is a form of dishonesty. An act of silence that is intended to cause another person to believe something that isn't true is deception. Our rationality often is the cause of our dishonesty. We want to maximize our payoffs. The higher the reward from being dishonest, the higher is the extent to which we engage ourselves to dishonest means.

We are vulnerable. We are vulnerable to attack, criticism, and temptation. Our vulnerability and susceptibility are subjective, and depend upon our cognitive abilities, personality, and social background. The 'upside' to vulnerability could be an asset. It is said that vulnerability has inherent goodness embedded in it. We try to balance our desire for personal gain with our willingness to be good. It is not wise to take advantage of someone's vulnerability, because once the situation changes, the vulnerable prey are no longer the defenceless prey. The vulnerable prey doesn't forget what his predator did to him. He waits for his sunny days to return, and when that happens, he doesn't lose the opportunity to pounce on the predator. And never forget that when a vulnerable acquires power, he is no less ruthless than a monster.

Like truth, the concept of wisdom is changing over time. The goal of wisdom is to comprehend the deeper meaning of the known facts, and for that a combination of cognition, self-reflection and openness are necessary. Wise look at events from different perspectives. They can weed out noise from the pattern. A truly wise can comprehend the limits of knowledge, as well as of wisdom.

Bertrand Russell said, "If knowledge continues to increase, the world will need wisdom in the future even more than it does now." Wisdom comes from experience. Wise have the ability to intuit the options before they become problems. As one wisdom researcher says, "One can have theoretical knowledge without any corresponding transformation of one's personal being. But one cannot have wisdom without being wise." The difference between a wise and a clever person, according to a Jewish saying, is that a clever can extricate him from a situation into which the wise would never have gotten him.

Wisdom is understood at the experiential level; it cannot necessarily be found in what a person says, but is expressed through an individual's personality and conduct in life. Wisdom is not necessarily conveyed through the content of a statement, but through the way the statement is delivered. Though wisdom is learned more easily than it is taught, but unless taught, it is learned the hard way.

We supposedly 'know' much more than our ancestors knew. But we have forgotten many things that our ancestors remembered. We can cure the dreaded cancer but we are forgetting the ways to lead a contented life. Wisdom does not automatically grow with age, but the association between wisdom and age is potentially positive. The biological "hardware" of the mind deteriorates with age, whereas the capability of "software" of the mind has the potential to increase with age.

The Mindset

Ellen Langer (7) conducted an experiment on elderly nursing home residents. One group was encouraged to live more fully; they were allowed to make more decisions for themselves. The second (control) group was not allowed to make their own decisions. For example, the first group was given houseplants, and was asked to take care of them, whereas, the second group was told that the nursing staff would care for them. A year-and-a-half later, Langer found that members of the first group were more cheerful, active, and alert. How did this happen? Langer explains, "the results were due to the power of making choices and the increased personal control it affords." She says that making choices results in mindfulness. We can change our physical health by changing our minds.

In another experiment, Langer and her colleagues sent two groups of men in their 70s and 80s to spend a week in an old monastery; "The right spot needed to seem timeless, with few modern conveniences." She wanted to find out "If we put the mind back 20 years, would the body reflect this change?" The first group was asked to pretend as if they were young men. The second group was told to stay in the present, and simply reminisce about the time when they were 20 years younger. For the first group, the clock was turned back, for the second group, it was not. The first group was asked to write a brief autobiography in the present tensc, the second group wrote their biography in the past tense. Before and after the experiment, both groups took a series of cognitive and physical tests. Langer noticed a change in behaviour and attitude in both the groups before and after the experiment. She found participants 'younger' in many respects. Both groups were stronger and more flexible; for example, both groups came out of the experience with their hearing and their memory improved. The first group (as if they were actually young) showed significantly more improvement.

Langer's experiments indicated that mind has enormous control over the body. Langer believes that our fixed ideas, internalised in childhood, can affect the way we age. Context matters; "I can see a candy bar from a great distance when I'm hungry." Her advice is to keep our mind open to possibilities because the power of possibility is huge. Mindfulness is a process of actively noticing new things, relinquishing preconceived mindsets, and then acting on the new observations. Mindlessness, on the other hand, blinds us to new possibilities. Langer argues that our mindless decisions can have drastic effects on our physical well-being. She says, mindful health is not about how we should eat right, exercise, or follow new age medicine, but about the need to free ourselves from constricting mindsets, and the limits they place on our health and well-being.

Our mindset has two forms: 'Fixed' and 'Growth'. According to Carol Dweck (8), those with fixed mindset have static intelligence and creative ability. A growth mindset, on the other hand, works on the premise that existing abilities can be extended. Dweck believes that the view one adopts profoundly affects one's way of life. The fixed mindset people are always in a hurry to prove their worth. The growth mindset people believe in the philosophy that everyone can change and grow through application and experience. What it means is that true potential is unknown, and it can be achieved through "passion, toil, and training." It says that deficiencies can be overcome. In the growth mindset, imperfections are not shameful. Its 'ideal' is not 'instant, perfect, and perpetual compatibility'. Dweck writes, "Just as there are no great achievements without setbacks, there are no great relationships without conflicts and problems along the way."

Our fearful mindset governs many of our behavioural practices. Fear is a natural part of human psyche. We have our own fears and we evolve our own ways to deal with it. We experience many kinds of fear: fear of death, fear of the unknown, fear of being alone, fear of the future, fear of failure, and so on. Fear also has the key to our survival. If we were not fearful, we wouldn't have survived. Evolution says, only those people survived who feared the right things at the right time. We have the tendency to anticipate. Anticipating a fearful stimulus can provoke the same response as and when it actually happens, and thus can be beneficial for dealing with the threat. Experiencing fear is a learned activity. Conditioned fear prepares us to learn about predicting adverse events. The obstacles to survival prepare our brain to deal with threats. Often it is hard to forget threats. The emotional impact of fearful events often continue for long. Accidents haunt us for a long time; driving at night seems dreadful if the accident had occurred during the night. The fearful memories easily return and are hard to shake off. Fear is frequently related to the escape and avoidance behaviour. When afraid we tend to fight, freeze, or flee. In the split seconds our adrenaline surges, eyes widen, heart rate increases, breathing quickens, stomach wrenches, palms moisten, and time slows down. These are the signs of fear.

Chaos scientists say that our brain works in unpredictable and random ways. They have used the concept of "self-organised criticality" to explain chaotic brain behaviour. Self-organised critical phenomena are driven by their intrinsic dynamic systems to reach a critical state, independent of the value of any control parameter. The perfect example of a self-organised critical system is a sand pile. As grains build up, the pile grows in a predictable way up to a certain point. Then suddenly the pile collapses. Though unpredictable, the overall individual distribution of sand is regular. The state of self-organised criticality lies right on the boundary between stable, orderly behaviour and the unpredictable chaotic world.

Neuroscientists say that our brains work on 'forest on fire' mode. In the forest fire, one burning tree sets alight another one. That is why whole forests don't catch fire all at once. Experiments have shown that single neuron can trigger fire to the neighbouring neuron, causing an avalanche of activity that can propagate across the network of brain cells. This results in alternating periods of quiescence and activity (much like the build-up and collapse of a sand pile). The brain often synchronises large groups of neurons to fire at the same frequency. This 'phase locking' process allows communication of different 'task forces' of neurons among themselves, without interference from others. During the process there is always a possibility of neurons firing out of sync due to interspersing of the stable periods of phase-locking with unstable periods.

A somewhat chaotic brain helps us, say neuroscientists, to cope up with our body's increasing demands. A chaotic activity is integral to learning and is believed to create new abilities.

Chaotic brains can have ordered thoughts. Mark Twain rightly said, "It usually takes more than three weeks to prepare a good impromptu speech."

A Corrupt Mind

A corrupt mind misuses the fertility of the mind, and as they say, a fertile but a vulnerable mind with no values is the ugliest mind. Corrupt mind is prime example of ugly mind. We are not born corrupt; we become corrupt. Corrupt wants to win at any cost. His overconfidence in his abilities makes him to adopt unjust, immoral and faulty ways. A corrupt is impulsive and his thinking is prejudiced. To fight corruption, more than the structural and legal reforms, moral renovation of the self is necessary. We need to develop innate values so as to keep us moral even in the most amoral situations. We need to elevate spiritual awareness in us. We need to refurbish ourselves with the lessons of trustworthiness, compassion, forbearance, generosity, humility, and courage. Corruption is generally associated with moral decadence. Its modus operandi is infinite. It can take the form of misplaced justice, misuse of authority, manipulation of public money, use of unfair means, fabrication of evidence, manipulation of data, use of or provision of banned substances to enhance performance, among others. Motivations for corruption may include economic gain, status, power, sexual gratification, etc. Measure of corruption subjective and depends upon a person's ethical, moral, cultural, and religious beliefs. In spite of the fact that perception and extent of corruption vary greatly across the world, the general agreement is that corruption damages society, democracy, and economic progress.

Ivan Petrovich Pavlov (9) in his studies on dogs observed that they start salivating the moment they see food; salivation happen even in the presence of some kind of food stimulant, like the lab coat of a person who served the food. Seeing the lab coat the dogs thought their food was on the way. Pavlov received similar response with ringing bells; the dogs associated the sound of the bell with food. Pavlov's concept of 'conditioned learning' says that events that previously had no relation to a given reflex (such as a bell sound) could, through experience, trigger a reflex (salivation). This work suggests that positive reinforcement causes repetition of the behaviour pattern, and also repetition causes the reactions to become more developed over time. Another important offshoot of 'conditioned learning', called 'extinction', is that an established conditioned response (salivating in the case of the dogs) decreases in intensity if the conditioned stimulus (bell) is repeatedly presented without the unconditioned stimulus (food).

Something similar must be happening in the minds of the corrupt. The corrupt becomes bolder with repeated stimulus. After tasting and liking the 'blood', and if he is not caught, the corrupt becomes doubly aggressive the next time. The corrupt can do anything (like lying, bribing, killing) to fulfil his desire. If not stopped, the corrupt becomes a bloodsucker. The problem is that the stimulus-response behaviour is so spontaneous and subconscious that it escapes the thought process of the corrupt. Pavlov's dogs tell us that if positive reinforcement works, so do negative enforcements. One of the positive reinforcements for the corrupt is power to prevail over others. To offset the positive reinforcement one would need negative reinforcement. Disqualification from holding a public office is one of the desirable negative reinforcements.

People will abuse power if we let them do so. The least we can do is to identify, isolate, and avoid the corrupt. In the 'corrupt space' there is need for fairness.

A Liar's Mind

To maintain a liar's mindset is a stressful activity. Defending a lie or a liar is not easy. Lying can be as complex as truth is. Not everyone can become a good liar. A professional liar masters the art of 'telling lies, the whole lie, and nothing but lie' in a manner as if it is 'the truth, the whole truth, and nothing but the truth'. Lying often becomes inevitable as telling the truth is not always desirable. More often than not, it is difficult to know the complete truth. And, as someone said, there are only two ways of telling the complete truth — anonymously and posthumously.

Some are habitual liars; some lie even when there is no apparent gain. Some lie out of respect for others. Some of us lie to reinvent ourselves. Some feel elated by fooling others. It seems fantasists find the experience of lying very rewarding. They feel it is one way to impress people. 'Chronic feeling of emptiness' drives some to lying. Accuracy and sincerity are the two virtues of truth. "People who claim we should never lie, not only neglect the second (sincerity), they also have an impoverished understanding of the first (accuracy)".

We value truth. Lying should be an exception, and therefore should require special justification. We withhold 'naked truths' in many situations. The idea that one should always say what one truly believes is narcissistic nonsense, argues a sociologist. His advice: one should say what needs to be said in a given situation. In many situations we take the umbrage of untruth. There is always more than one way to give a truthful description of an event. One can describe a situation highlighting a particular aspect that suits one's own perspective. Without lying one can send across a wrong picture. If the intention is to deceive, in spite of no-false statement, a statement can be made with 'plenty of economy with the truth'.

There are some neurological patients who construct false answers while genuinely believing that they are telling the truth. There was a man, a stroke victim. He described a conference that he never attended. He constructed false answers and thought he was not lying. There was a paralytic patient. She won't admit it. She would say she has arthritis. She too was not lying. 'Confabulation' is genuine lying. It is genuine because it "lacks the intent to deceive". It is an attempt to fill in memory gaps by fabricating information or details. In confabulation, first a false response is created, and then the patient fails to recognise its falsity. A normal person, on the other hand, has the capacity to correct the falsity, if he has created it.

Our memory is reconstructive. During the reconstruction process, the memory is pieced together from fragments. While doing so errors creep in. The errors could be due to biases and expectations. The error can also be due to the damage in frontal lobe (due to tumours, head injuries or ruptured arteries). Our memories are usually reliable as the errors that creep in are small. In extreme cases, memories can be completely false. Confabulation is unintentional false memory. It occurs due to the confusion of imagination with memory, or the confused application of true memories. It involves the absence of doubt about something one should doubt. We love to tell stories, more so, if the story is about ourselves. If we get an opportunity to tell others how honest, ethical, successful and interesting we are, we try to seize it. "Perhaps confabulation arises in part from this natural inclination toward telling stories about ourselves."

We are often blind to many things that surround us. We fail to notice things that are right in front of us. Our brain is 'selectively selective'. We love prioritizing few things, and neglecting few other things. We make deliberate mistakes, because we don't want to confront problems that are likely to create further problems for us. We have an unconscious tendency to simplify the world. We are afraid of being misunderstood. We don't like to be scrutinised by others, nor do we like to scrutinise others, particularly the powerful. We prefer keeping our eyes closed as per our convenience. We love transferring our responsibilities to others. We like to convince ourselves that the problem would resolve by itself, and any kind of intervention is not needed to resolve the problem. This problem is more problematic in small groups, where one of our prime motives is to exchange niceties. As they say, "The cosier and the more close-knit the group, the less incentive you have to stir the waters." Our loyalty often tends to overprotect. When we don't want to discuss an issue we simply try not to discuss it. The matter is as simple as plotting points on a graph paper. The points that are not as they should be, either we ignore them, otherwise we should be willing to explain the deviations. It is easier to ignore than to explain. Why bother to find out if it was due to experimental error, or due to some new phenomena that was responsible for the deviation from the expected norms. Either way, it is additional work, so why bother.

We, the humans, are designed to make mistakes. We are subconsciously biased. We are overconfident of our own abilities, and that leads us to make mistakes. We make two kinds of mistakes when an undiscussable issue is raised: one, we fix blame in a way that escalates stress and conflict, and two, we use ambiguous language that enables people to avoid the problem. When companies have a culture in which managers are more interested in hiding problems than solving them, very little one can do.

Brain in Order and Disorder

Nothing is more important than a brain in order. On the other hand, nothing is worse than a disordered brain. "How to keep brain healthy" is a major neuroscience challenge. It is said that the secret of cognitive smartness lies, not only in the matter, but also in their arrangement. The human brain is more compact and more organised compared to other big-brained creatures. The thicker insulation around nerves in the human brain is responsible for developing faster information processing capability.

As we age, our cognitive abilities (such as memory loss, speed to process information, reasoning ability) decline. If the rate of decline of cognitive functions is too high, it may lead to Alzheimer's disease, or some other form of dementia. We keep our brain healthy by protecting, restoring, and enhancing its functioning. Healthy life style can prevent one from the drudgery of many devastating diseases, including Alzheimer's.

It means, chronologically old but biologically young people can have nearly normal long life. Nutritious diet, physical fitness, social engagement, and mentally stimulating activities are potential factors that help in reducing the risk of cognitive decline. Scientists are, therefore, looking at many possible interventions, such as cardiovascular and diabetes treatments, antioxidants, immunization therapy, cognitive training, and physical activity to slow down, delay, or prevent cognitive decline.

Healthy ageing has many dimensions. Some requisites are: avoidance of disease and disability, maintenance of high physical and cognitive function, sustained engagement in social and productive activities. It is believed that early years of childhood can shape many adult outcomes. Economics shows another dimension of ageing. Economist Robert Fogel believes that increase in longevity and in health is too rapid to have been caused by genetic or evolutionary changes. He believes, it is due to better control of our physical environment. According to him, factors like nutritional intake, economic growth, investments in health and greater scientific knowledge are important factors that improve longevity.

Ageing is biological destiny, and not a disease. Wise say, growing old is a natural process, and should be treated as such. More crucial than extending life span is extending the health span.

Questioning Mindset

We have a questioning mindset. Our questions reflect our ability of reasoning, understanding, and learning. They also reflect our hollowness. We ask close-ended convergent questions, as well as, open-ended divergent questions. Close-ended questions are 'saturated'; divergent questions are 'unsaturated'. Questions can be 'deeply satisfying', and also 'deeply troubling'. Questions often help us to move forward by taking us away from our comfort zone. Some questions have both 'yes' and 'no' answer. Some questions have no answer.

Milan Kundera thus articulates, "It is questions with no answers that set the limit of human possibilities, describe the boundaries of human existence."

The ability to form a good question is one of the key cognitive abilities that separates one from the other. "The desire to ask a question shows a higher level of thought, one that accepts that your own knowledge of a situation isn't complete or perfect", writes Phil McKinney (10). The ability to think inquisitively is one of our critical survival skills. Genius has a high level of cognitive disinhibition. They have the mental agility to process and use all the absorbed information in an organised manner. McKinney believes that anyone can develop and harness this power through the use of provocative questioning, and the discovery that follows. Often we ask questions, but don't expect any answer. Often we expect only approbation.

Socrates believed that the first step towards knowledge is the recognition of one's ignorance. Socratic questions are for probing assumptions, reasons and evidences about viewpoints and perspectives, and probing implications and consequences. They challenge beliefs about a subject, rather than responding with an answer that has been taught to be 'correct'. Socratic questions don't assume you are right or wrong. Their goal is to unveil thoughts and beliefs.

John Brockman, Editor, Edge (www.edge.org), has asked many interesting questions. Take a snapshot: What scientific term or concept ought to be more widely known? Some interesting responses are: Premortem (an effective tool to avert disaster before the disaster); Affordances ("the perceptual systems of any organism are designed to "pick up" the information that is relevant to its survival and ignore the rest"); Naïve Realism ("term explaining why we see most people other than ourselves as unintelligent or crazy"); Invariance ("Science is supposed to be about an objective world. Yet our observations are inherently subjective"); Optimization ("Incremental modification, followed by evaluation and readjustment, guides us to solutions that maximize a desired criterion"); Premature Optimization ("Why do the successful often fail to repeat their success? Because success is often the source of failure. Success is a form of optimization"); Deliberate Ignorance ("the wilful decision not to know the answer to a question of personal interest, even if the answer is free, that is, with no search costs"); Need For Closure ("desire for a firm answer to a question. When NFC becomes overwhelming, any answer, even a wrong one, is preferable to remaining in a state of confusion and doubt"); Antisocial Preferences ("willingness to make others worse off even when it comes at a cost to oneself"); Positive Illusions (ability to convince ourselves that the real is the ideal); Mental Emulation ("a way to simulate what you would expect to happen in a specific situation"); Relative Deprivation ("that idea that people feel disadvantaged when they lack the resources or opportunities of another person or social group").

An Ideal Mind

Sadi Carnot conceived a 100 per cent efficient engine. Mathematicians conceived an abstract concept that has no limit. Carnot's engine and infinity are the best examples of an ideal entity. Ideal is a goal post that can't be reached. One may ask - Does the idea of ideal makes sense in a non-ideal world? A related question can be - If a concept can't be achieved, should we try to achieve it? I think we should. We need a goal post, even an imaginary one, to reach there. If there is no such goal post, there wouldn't be any progress.

We are subjective idealists. We follow our own brand of idealism. The problem is that our purposes and values are dynamic, and thus our idealism is provisional and tentative. 'Ideal' is like a thought experiment that can be thought of, but can't actually be performed. A thought experiment is performed, nevertheless, because of its useful implications.

Man conceived an omnipotent and omnipresent entity. God, no one has seen, some pretend to have seen, while some assume to have seen him. For some god is equal to zero, and for some god is equal to infinity. It means, god is zero as well as infinity, depending upon who the mathematician is. One of its interpretations could be god exists as well as doesn't exist. It is we who decide if god exists or doesn't exist. It is we who creates god as per our convenience. We procreate him when we need him. He doesn't exist once our needs are over. We remove him from our want list as and when he is not needed. Our wants and needs decide the fate of god.

God is someone who has no eye witness. No one has seen his footprints. Some think he is the creator of stars, galaxies, human beings. He arose out of an illusion, and can create something out of nothing. He arose because he was needed to dampen the abundance of chaos. He was needed to keep the chaos at bay. The question many are asking is - Why God can't be less omniscient and omnipotent? Why can't God be less god-like and more human-like?

Some think God is 'belief in hope beyond reason'. "There seems an inherent human drive to believe in something transcendent, unfathomable and otherworldly, something beyond the reach or understanding of science," writes Robin Maratz Henig (11). We may doubt the existence of God, but we act as if we believe in something, by whatever name we may call that thing. It is as difficult to prove the existence of god, as his non-existence.

It is said that "God created man in his own image." It is also said that "men create the gods after their own image." It is like, for our survival we fit into the environment, and also we create the environment, as per our survival needs. The number of followers of god is ever increasing. It is, perhaps, because religion serves a distinctly human need for social interaction, and has the potential to offer comfort and peace of mind in the world full of uncertainties and unknowns. There are always good reasons to believe or disbelieve in God. It is a highly complex issue.

When Albert Einstein was asked, do you believe in god, he beautifully summed up his predicament: "We are in the position of a little child entering a huge library filled with books in many languages. The child knows someone must have written those books. It does not know how. It does not understand the languages in which they are written. The child dimly suspects a mysterious order in the arrangement of the books but doesn't know what it is. That, it seems to me, is the attitude of even the most intelligent human being toward god. We see the universe marvellously arranged and obeying certain laws but only dimly understand these laws." Einstein's God did not reward or punish. His God was a 'conceivable mystery'. He thought that finer speculations in the realm of science can't be understood without a deep religious feeling. Writes Einstein, "A knowledge of the existence of something we cannot penetrate, our perceptions of the profoundest reason and the most radiant beauty, which only in their most primitive forms are accessible to our minds, it is this knowledge and this emotion that constitute true religiosity; in this sense, and in this alone, I am a deeply religious man."

We create God in our own ways. Our expectations from god define our God. Rabindranath Tagore's god is unknown and unknowable, beyond one's reach. His god is always with man, in his happiness and in his pain. His god possesses all the metaphysical, moral and causal attributes. His god has the knowledge of all the things. His god is integral to man and nature. His god is above the pairs of opposites, of beauty and ugliness, truth and error, good and evil. His god can't be seen; he can only be realized. His god doesn't live in far-off heaven. His god is both individual and universal. For Rabindranath, "Without the world God would be phantasm; without God the world would be chaos." His god is available in countless forms. His god is the inspirer of his songs. Rabindranath could hear his silent steps. Tagore's god needs man's love, as much as man needs god's love. His god is infinite, like a sky. His god is beyond touch, but can be touched. His god can be found in lonely streets, walking alone. Often Tagore could hear in the dark alleys the songs of righteousness.

Leon Lederman (12) writes, "In the very beginning, there was a void, a nothingness containing no space, no time, no matter, no light, no sound. Yet the laws of nature were in place, and this curious vacuum held potential. Only god knows what happened at the very beginning?"

Simple logic says that something has a better chance of survival, if it is advantageous to us. One of its corollaries is that belief in something can sustain, only if that thing proves its worth. The belief system will fizzle out if it is not worth believing. God, for example, would not have survived, had he not proven his worth.

Psychologists say, the experience of the illusion of God is universal. The illusion is that someone up there is constantly watching us, and is also concerned about our moral lives. When we know someone is watching us, we tend to behave differently. 'Someone is watching us' is our basic nature. Even devoted atheists possess this nature. Our belief in God shall continue, as long as he is useful to us.

Studies indicate that supernatural beliefs promote altruistic behaviour and adherence to social norms. Researchers say that 'belief networks' operate across the brain and invoke spirituality. Some even say, atheism is probably the unnatural way to be. It is also said that religion is our survival instinct, because it brings us closer to the other.

It is quite possible that God is an evolved projection of our understanding of what God should be. In his book The Singularity Is Near, Ray Kurzweil, writes: "Evolution moves toward greater complexity, greater elegance, greater knowledge, greater intelligence, greater beauty, greater creativity, and greater levels of subtle attributes such as love." He then adds, "God is likewise described as all of these qualities, only without limitation. . . . So evolution moves inexorably toward this conception of God, although never quite reaching this ideal."

God is a virtual reality, as mind is. Virtual and augmented reality extends human awareness and allows us to see the big picture. Some say it is more real than the real reality. Both real and virtual realities are limited by our brain. The point is how real our reality is. Can we evolve our brain to such an extent that it becomes God's brain; only God's brain can go to unimagined places, both virtual and real.

Our brain defines us. Our brain can see meaning even where there is no meaning. We make images of god in our mind, and then project it as if it is real. Our mind has the ability to conjure up the imaginary. Everyone has his own god, because everyone has his own brain. Our realities vary.

Why would the human imagination invent a god? It is because the human mind has the power of imagination, and it wants to reach beyond the real.

Are we trapped in God's video game, or God is trapped in our video game, our mind only can tell.

Indian Mindset

Mindset is a reflection of cultural identity. Mindset reflects its attitude towards work, negotiating style, decision making, problem solving etc. An eastern mindset is circular. A western mindset is linear. The easterner believes that the world is constantly changing the way one moves along a circle and is thus not so easily controllable, writes social psychologist Richard Nisbett (13). A westerner thinks sequentially and logically and thus the changes are predictable and controllable, as in a line. A westerner straight away comes to the point, because his objective is to forge a good deal. An easterner, on the other hand, comes to the point through a long and circuitous route, as his objective is to forge a long-term relationship. An easterner believes that there is no absolute truth but strives to reach it. For an easterner, collective is more important than the individual, and, therefore, an easterner is 'relationship focused'. A westerner believes in absolute truth, and believes in the power of individual to influence the environment, and is 'object focused'. For an easterner, 'life is a mystery to be unravelled'. For a westerner, 'life is a problem to be solved'. The easterner believes in discussions to arrive at a harmonious solution. Westerner's approach is debate.

Biologist T Motokawa's interesting article, Sushi Science and Hamburger Science, highlights the attitudes of eastern and western mindsets. Motokowa says that when a westerner reads papers written by Japanese scientists, he often finds it difficult to understand. The reason is the difference in logic. "Western logic is quite clear: it has a structure in which each statement is tightly connected and linearly arranged to reach a conclusion. Japanese logic is not clear.... Japanese people talk about something and, without stating a conclusion, move the discussion to another topic. These two topics often have no logical connection, although they are related in the mind of Japanese people. What the Japanese are trying to do is to describe one fact from various points of view. Each view is connected by imagery to others, not by strict logic."

Every culture has few dominant traits. India is no exception. It is known for diversity. It has long cultural history. Strong stamp of legacy is imprinted on the mindset of its people. More deep in history one is, more time the mindset takes to change. Indian orientation, thinking, and attitude are subjective. How an Indian reacts to a particular situation depends on their nature, as well as nurture. This in some sense explains why our attitudes and approaches to understand common problems are different.

Bipin Chandra Pal (14) beautifully describes the Indian psyche. Indian psyche is "more transcendental than formal, more metaphysical than scientific, more imaginative than positive, more idealistic than realistic". Indian values are more intellectual than physical, internal than external, emotional than rational. When one walks a mile to meet a friend, he feels he has walked only a few steps. If he has to do unpleasant work, he will say he walked a mile, when, in reality, he walked only a few steps. When a friend meeting a friend after a few weeks says, I have not seen you for ages he really, neither exaggerates nor lies, but simply applies his own inner emotional standard to measure time.

Indians are too expressive of the way they talk and display their enjoyment, which a westerner may find too intrusive. Indians love bright coloured dresses. These dresses may seem loud and flashy to a westerner. An Indian advertising guru nicely sums up the situation with a film analogy: "While both Black and Iqbal are well-made films, Black is India trying to ape the west, while Iqbal is pure Indian creativity."

Indian mindset loves paradoxical situations. It likes to accept the new, but finds it difficult to reject the old. In the process, Indians get branded as "elusive, hypocritical and unreliable." They are highly sensitive, but their actions often lack the spirit of sensitivity. Fair play is as important to them as the foul play. There can't be just one tag line for Indian mindset. Some mindsets are highly optimistic and entrepreneurial; they can play the game whatever the rules are. To some mindsets opportunities make no dent. For small personal favours some can forego big gains.

Jai BP Sinha (15) thus describes Indian mindset: "Indians are both collectivists and individualists, hierarchically oriented while recognizing merit and quality, or resisting dominance, religious as well as secular and sexually indulgent, spiritual as well as materialists, excessively dependent but remarkably entrepreneurial, nonviolent by professed values but violent in behaviour, and comfortable in taking analytical, synthetic as well as intuitive approaches to reality. Such a coexistence of opposites often causes inaction, perfunctory action, feuds and infightings, but also equips them to be creative and jugaru (improvisers) through continuously aligning their thought and behaviour to the demands of a milieu." Often trends influence Indian mindset. Often Indians are unaware that they are being influenced. Social influence sometimes attracts and leads Indians to do thing as others do, but in other times it repels to do something different.

Individuals from independent cultures tend to value their autonomy, uniqueness, freedom, and right to self-expression; whereas individuals from interdependent cultures tend to prize social harmony, conformity, and adherence to group norms. "The mere fact that someone else is around can change how we behave," tells Jonah Berger (16). For things Indians are good at makes them perform better in the presence of others. Indians perform worse in the presence of others doing the acts Indians are not good at.

Mark Twain wrote in 1897 about India, "the land of dreams and romance, of fabulous wealth and fabulous poverty, ...of tigers and elephants-the country of a hundred nations and a hundred tongues, of a thousand religions, and two million gods, ...grandmother of legend, great grandmother of tradition..."

Will Durant (17) writes in 1931, "And though I have prepared myself with the careful study of a hundred volumes, this has all the more convinced me that my knowledge is trifling and fragmentary in the face of a civilization five thousand years old, endlessly rich in philosophy, literature, religion and art, and infinitely appealing in its ruined grandeur and its weaponless struggle for liberty."

Mother India is in many ways the mother of us all, Durant writes. "Nearly every kind of manufacture or product known to the civilized world-nearly every kind of creation of Man's brain and hand, existing anywhere, and prized either for its utility or beauty- had long, long been produced in India. India was a far greater industrial and manufacturing nation than any in Europe or than any other in Asia. Her textile goods- the fine products of her looms, in cotton, wool, linen and silk-were famous over the civilized world; so were her exquisite jewellery and her precious stones cut in every lovely form; so were her pottery, porcelains, ceramics of every kind, quality, colour and beautiful shape; so were her fine works in metal-iron, steel, silver and gold. She had great architecture. She had great engineering works. She had great merchants, great businessmen, great bankers and financiers. Not only was she the greatest ship-building nation, but she had great commerce and trade by land and sea which extended to all known civilized countries. Such was the India which the British found when they came."

India is a functioning democracy. Chaos and conflict is part of the democracy. "Chaos and conflict, arising out of diversities of languages, religions, ethnicities, have not broken up the country," writes Jean Dreze and Amartya Sen (18). Dreze and Sen, however, are not sure of the glory of the much talked-about achievements in new India. One of the reasons of 'uncertain glory' is "a lack of serious involvement in the diagnosis of insignificant injustices and inefficiencies in the economic and social lives of people." Their list of shortcoming includes unequal income distribution and stagnant real wages, public revenue not properly used to expand social and physical infrastructure, inadequate social services to large part of the populace, failure in governance and organization of power sector, water supply, drainage, garbage disposal, public transport, accountability and corruption.

In The Argumentative Indian (19), Amartya Sen writes, "Our ability to talk at some length is not a new habit. Our epics Ramayana and Mahabharata are its testimony. These proceed from stories to stories woven around their principal tales, and are engagingly full of dialogues, dilemmas and alternative perspectives. And we encounter masses of arguments and counterarguments spread over incessant debates and disputations." India's argumentative traditions are still alive and have continued to have their influence. Sen thinks, 'seeing Indian traditions as overwhelmingly religious, or deeply anti-scientific, or exclusively hierarchical, or fundamentally unsceptical involves significant oversimplification of India's past and present."

India: its history lies in obscure ruins, not in museums, its religion proliferate in everyday life, not in grand organized churches, and its food is best had at homes, not in restaurants. Its heat is severe, its rain unending," is how Pankaj Mishra (20) sums up India in India in mind.

India was presented as a package of all ills, obviously with a vested interest. Pawan K Varma (21) writes, "the country was seen as irrevocably fragmented or spiritually transcendent, hugely ungovernable or simplistically self-reliant, venal beyond redemption or blissfully unmaterialistic, impossibly opaque or wonderfully ancient and revealing." Indian mind needs better interpretation in the eyes of the world, thinks Varma. It has a strong legacy. Projecting things Indian is no longer a shame. India is now an 'awakened elephant'. People now come not only to see the amazing monuments but also to participate in joint ventures that are truly 'Joint' in nature. India is no longer opaque. Indians themselves have done image build-up, as they found it the only way. India is becoming so important that it can't be neglected anymore. Uncertainty no longer defines India. Indian can retain hope, and has resilience to adversity. India, a land of contradictions, a mixed bag of tradition and modernity; one can't live without the other.

Future of Mind

Are the 'soft and biological' humans perfect? Or, more perfect 'hard, digital and almost inconceivably powerful' humans are yet to evolve. Is our blueprint, like many other blueprints, likely to obliterate completely? Hopefully not!

"Extrapolating the trajectory of our own current technological evolution suggests that with enough computational sophistication on hand, the capacity and capability of our biological minds and bodies could become less and less attractive," writes Caleb Schrf (22).

Michio Kaku (23) says, in the future, it will be possible to read brain by combining the latest scanning technology with pattern recognition software. Kaku envisages a mind that can videotape dreams. Our consciousness can be downloaded onto machines. There would be the possibility of transporting thoughts and emotions through the "internet of the brain". Kaku is optimistic about the future of mind. "Perhaps one day the mind will not only be free of its material body, it will also be able to explore the universe as a being of pure energy. The idea that consciousness will one day be free to roam the stars is the ultimate dream. As incredible as it may sound, this is well within the laws of physics."

Kaku says, mind is a hard engineering problem, and since fundamental laws of engineering are already known, it will be easier to understand and manipulate the 'computer of meat'. By mapping the 'connectome', Kaku imagines, it should be possible to reverse-engineer each and every person's brain. Using the connectome one can download oneself into a machine. When that happens, mind can live, as long as that machine lives.

There is worldwide interest in reverse engineering the brain. Major groups that are involved in the work have varied aims. Some aim to simulate the brain electronically on computers. Some aim to map the neurons of the brain. Some aim to decipher the genes that control brain development. These developments raise an important issue: Are we merely the sum of our brain's connections? In the absence of a good working model to understand consciousness, should we envisage the kind of future of the mind that is being envisaged? Can machines ever surpass human intelligence?

Computers, like brain cells, are interconnected. It is thus logical to assume that one day computers would work the same way human brain works. But, both neuroscientists and computer scientists know that human brain is different from a computer in number of ways. They know that synapses are far more complex than electrical logic gates. They know that inter-neuronal connections in human brain are much more than in any computer. The brain is a massively parallel machine; computers are modular and serial. The brain uses content-addressable memory, computers byte-addressable.

Brain uses evolution; adults have far fewer synapses than many random synapses the toddlers have. There is no central clock in the brain, and its processing speed is not fixed. The human brain constantly rewires and self-organises itself. No hardware/software distinction can be made with respect to the brain; the mind emerges directly from the brain, and changes in the mind are always accompanied by changes in the brain. Brain has body at its disposal so that it can 'offload' its memory requirements to the environment in which it exists. Though the human brain has the capacity of learning things faster than computers, it is slower than computer in many functions, like multitasking capability, and mathematically involved processes are better with computers. The brain is capable of imagination, and is far superior to computers in matters related to common sense.

The developments in artificial intelligence, robotics, and machine learning raise mind-blowing, as well as mind-frightening, possibilities. The question that is most bothersome is - Should machines surpass human intelligence? The related question is - Can machines ever surpass human intelligence? A related counter question is - Since computers are creating so much mess in the information cyberspace, should we not divert our attention to clear the mess, rather than creating additional mess? Should we not control 'out of control' cyber pollution, in addition to installing better security guards?

David Gelertner (24) believes that no computer will be able to think like man unless it can 'free-associate'. By 'free associate' he means, "When you stop work for a moment, look out the window and let your mind wander, you are still thinking. Your mind is still at work. This sort of free association is an important part of human thought". Gelertner also says that no computer will be able to think like we think, unless it can 'hallucinate'. We hallucinate when we fall asleep and dream, and in the hallucinated state our mind redefines reality for us; outside reality disappears, inside thinking remains. Gelertner contends that level of 'alertness' is basic to human thought. In the low alert state, our thoughts tend to move by themselves with no conscious direction. In this state of free association, each new thought resembles or overlaps or somehow connects to the previous thought. With fall in our alertness we lose contact with external reality. Eventually we sleep and dream. In the hallucinated state the thinker and his thought stream are not separate. The thinker inhabits his thoughts. No computer will be able to think like a man unless it, too, can inhabit its thoughts, and can disappear into its mind.

"Brains are undoubtedly somewhat computer-like – computers, after all, were invented to perform brain-like functions – but brains are also much more than bundles of wiry neurons and the electrical impulses they are famous for propagating," writes Alan Jasanoff (25). But then he says, "not every brain cell is sacred". The logic behind his statement is the "apparent retention of memory after removal of half of the brain, ... and by the retention of the child's personality and sense of humour". In other words, sheer number of cells in the human brain is unlikely to explain its extraordinary capabilities. Jasanoff is of the view that there is no causal or conceptual boundary between the brain and its surroundings. "The integrated involvement of brain, body and environment is precisely what makes having a biological mind different from having a soul, and the implications of this difference are tremendous."

Brain can't be looked in isolation. It is an integral part of the society we live in and the environment that surrounds us. Our culture and resources shapes it. Jasanoff wants us to believe that we are much more than our brains.

2

Science Refines the Spirit of Life

Science has individual, social, and institutional dimensions. Because of the social relevance of science, the dissemination of scientific information is crucial for its progress and contemporaneity. A scientist observes and then tries to make sense of that observation. She/he collects data, develops theories, uses hypothesis to understand and explain a thing, or a phenomenon, and then tries to find patterns emanating from her/his observations. Scientific knowledge keeps on changing.

Science is a blend of logic and imagination; therefore, there is always a possibility of disagreement and mismatch between them. A hypothesis is often untestable. Such untestable hypothesis are often rejected by the scientific community. "Science never pursues the illusory aim of making its answers final, or even probable. Its advance is, rather, towards the infinite yet attainable aim of ever discovering new, deeper, and more general problems, and of subjecting its ever tentative answers to ever renewed and ever more rigorous tests," writes Karl Popper (1).

A scientist formulates, corrects, or discards the theories. The job of a scientist is to discriminate false idols from the real things. Science has built-in error correcting machinery, writes Carl Sagan (2). The error bars are insistent reminders to the fact that no knowledge is complete or perfect. It is used to exercise self-criticism. Each generation of scientists reduces the error bars a little. When we are self-indulgent and uncritical, when we confuse hopes with facts, we slide into pseudoscience and superstition. Sagan believed in maintaining an essential balance between the two attitudes: openness to new ideas, no matter how counter-intuitive they may be, and the most ruthlessly sceptical scrutiny of all ideas, old and new.

Rick Borchelt (3) sees in George Seurat's painting - A Sunday in the Park on the Island of La Grande Jatte - a perfect metaphor for science communication. Scientists describe the dots. Members of the public want to understand the picture. Scientists not only connect the dots, but also communicate with the general public in a way that maintains their trust. It is the responsibility of the scientist not to hype or oversell science.

The job of a scientist is to discover patterns, use logic and imagination to establish a fact. The metaphor of 'gardener of ideas' has been used for scientists. Some say, "Such intellectual gardeners usually pay little attention to toads or other uninvited creatures residing among them, unless those creatures begin to munch or trample the plants."

A scientist is described as a hybrid of monks, artists, and practical contractors engaged in building a cathedral of unlimited scope. Outsiders may be concerned that the construction is expensive, and that they are not sufficiently involved in the process. The present day monks, however, have become worldlier. They are opting science to make a living.

Nature Lab

Experiments are important part of a learning process. We conduct experiments to get some data. If the data meet our prior expectations, we call it a good experiment. If it doesn't, we try to revise our prior understanding.

Life can be described as a sum of experiments, but it is difficult to conduct controlled experiments in most aspects of our lives. We can't plan our lives meticulously. It is impossible to anticipate all random signals that we are supposed to interact with in all areas of our activities. In the areas where prediction is possible, detailed planning can protect us from going out of control. When we live life as an experiment, we should be willing to take the risks and acknowledge failure.

Nature plays a significant role in our life. Nature is the best lab for conducting the experiments of life. It can offer so much. Nature teaches us to enjoy the world amidst various colours and contradictions. Amidst chaos, it teaches us to enjoy the experiences of the quiet. The best thing nature teaches us is that everything has a purpose. It encourages curiosity. It inspires us to observe things that may not have immediate interest for us. Nature may not provide comfort in the usual sense, but it does provide comfort in the deepest sense. It has calming effect; it keeps in track our wandering mind. It teaches us to let go when we need to. It teaches us gratefulness. It tells us that we are here to play a small part in a much larger scheme of things. It tells us that our part may be small, but it is useful in the overall scheme of things.

Nature tells us that faster one diversifies, better is the evolution. It tells us that greater the capability to adapt, better are the chances of winning. Evolutionary biologists say, birds are evolutionary winners, because they have diversified much more than they were expected to. Crocodiles, on the other hand, are evolutionary losers, because they have diversified much less than they were expected to.

Nature has emotions. It has feelings. It is full of diverse organisms. Plants feel pleasure as well as pain. They respond to external stimuli. Their movement is restricted, but they move when the need arises. Plants can protect themselves from wounds. Research suggests that plants can remember and react to information contained in light, and can use the information to immunise themselves against diseases.

Jagadish Chandra Bose believed in the fundamental unity of all lives, but he believed that all life forms don't have the same level of consciousness. Like Bose, Barbara McClintock believed in the oneness of things. In order to love plants, she used to say, one must have feeling for the plant. One must have "feeling for the organism," says Evelyn Fox Keller (4). By 'feeling' she meant that one must be able to hear what the organism is saying; when it wants to come near you, let it come; one must recognise that different plants, like different humans, need different treatments. Mere seeding is not enough for a plant to grow. Feeling for the organism means to go along with the plant, as it grows. McClintock says that our scientific intelligence is not sufficient to understand the ingenuity of nature. Plants have the capability of doing much more than our 'scientific intelligence' can perceive.

Nature is lawful, and it expects its habitats to be lawful. To understand nature one needs to practice and follow good science. "Good science cannot proceed without a deep emotional investment on the part of the scientist," believes Keller. Without emotional investment, spending long lonely hours in the lab doesn't make sense. In the organisation of nature there is little scope for randomness. If there is breakdown, it is due to the inadequacy in our understanding nature's actual order.

Scientific Collaboration

More and more people are realising the importance of relatedness among unrelated things; the more unrelated the elements are, the more radical and innovative is the synthesis. In the fast changing world, collaboration among diverse disciplines is the norm. The knowledge sharing is possible when basic values and understanding among the collaborators match. A conceptualist can collaborate with an experimentalist, but their roles and their approach to problem-solving are different. This is not to say that others' views should be accepted without critical examination, particularly when it comes from entirely diverse sources.

In scientific activities collaborators are sharing and complementing conceptual and experimental approaches. In the practice of art personal taste, vision, and style of expression matters. These shared characteristics need better understanding. There is so much truth in the beautiful observation of Murray Gellman, "What is especially striking and remarkable is that in fundamental physics a beautiful or elegant theory is more likely to be right than a theory that is inelegant."

Generally, scientists are known to be serious types. Artists, on the other hand, are considered more playful. Does playful approach diminish seriousness? Is science only a mechanical endeavour, because its emphasis is more on norms and methodologies? When we think of scientists do we expect them to be lonely scholars working in a monk-like mode in the corner of a room, or we see in the "hunches of a scientist the echoes of intuition of an artist."

In a party, you meet a 'cool' scientist. This person is fun-loving, loves music and films, wears jeans, can play guitar, and can sing. Beside these things she/he can do many other things that are normally not associated with science. Would you be disappointed after meeting this person, as your expectations were so different? Would you wonder how a fun-loving and animated person can be a scientist? Would you wonder how 'cool' people can do really interesting 'hot' stuff in the labs? I presume you will not be disappointed. Cool scientists are now becoming the norms of the society. Someone rightly pointed out that there is a huge divide between how people perceive of what happens in science and what actually happens.

One of the purposes of interaction of a scientist and artist is to break the barriers. More and more of either kind are now trying to break the barrier. More and more scientists are excited to collaborate with artists. The world of the academics is changing very fast, and with the changes, we are observing new attitudes of the players. The monotony of a quiet life may stimulate the creative mind, but there is also a realisation that collaboration helps. As a result, academics are becoming more interactive.

As someone put it, teams trump soloists when it comes to scientific output and impact. Increasingly large number of people are realising the importance of relatedness among unrelated things. The success of collaboration, however, will depend upon the existence of a shared body of knowledge and techniques. This knowledge sharing is possible when basic values among the collaborators match.

Collaborations work when the role of each contributor is seen and assessed in proper light. When two diverse thinkers collaborate on a project, collaboration should work, as the purpose of collaboration is to accommodate diverse thinking. This is not to say that one's view should be accepted without further examination, particularly when the views are coming from entirely diverse sources. To hold collaboration for long is not easy.

Collaborations in science are usually short-lived than collaborations in visual arts and humanities. In the sciences, collaborators share conceptual or experimental approaches, whereas in arts, it also includes personal tastes, values and the style of expression. The need for such deeper shared characteristics explains why there are fewer collaborations in arts, and the collaborations that do exist are long lasting, and often combined with personal partnerships, opines David Galenson and Clayne Pope (5).

Dean Keith Simonton (6) distinguishes creative scientists from creative artists on the basis of their training. According to Simonton, scientific creativity requires much more formal training than is required in artistic creativity. "Indeed, some studies have found a curvilinear inverted-U relation between artistic creativity and formal education level so that those with higher degrees are at a relative disadvantage." Simonton says that scientists are better students than artists. The third dimension of difference between an artist and a scientist is in the influence of mentors. "Although the creative development is enhanced by working under notable creators in the same domain, artistic talent is best nurtured by studying under a diversity of artists while scientific talent is better nourished by studying under just one."

Science Communication

Scientist can bring information, insights and analytical skills to bear on matters of public concern. In playing this advisory role, scientists go to great lengths to avoid bias—their own, as well as that of others. But in matters of public interest, scientists, like other people, become biased when their own personal, corporate, institutional, or community interests are at stake. There is a view prevalent in the 'those-who-matter' circle that "on issues outside their expertise, the opinion of scientists should enjoy no special credibility." This proposition is debatable. Isn't it an irony that non-scientists take major decisions related to the application of science in public life?

Albert Einstein said, "Most of the fundamental ideas of science are essentially simple, and may, as a rule, be expressed in a language comprehensible to everyone."

Educating the common man about science is one way to enrich and activate their thinking process. Scientists are expected to tell their stories simply and cogently. They are expected to reach beyond peer-reviewed publishing. It is generally believed that scientists are not effective communicators, if the communication is meant for non-scientists. Many scientists are reluctant to engage with people outside their own community, because they think this responsibility doesn't come under the purview of their 'job'.

Scientists are often branded arrogant, as they try to impose their theories on the common man. They are also branded as the "destroyers of the awe and wonder of nature." However, as Sagan argues, "Scientists do not seek to impose their needs and wants on Nature, but instead humbly interrogate nature and take seriously what they find." The 'ivory tower mentality' of scientists is a matter of concern for some. They say, scientists are only interested to understand the laws of nature. They are less concerned about the general human affairs. This perception was perhaps more in vogue in the past, when a scientific finding and its practical applications were well separated in time and space. This detachment enabled scientists to absolve themselves from the responsibility for the effects their findings might have on other groups in society. The situation now, however, is different.

Scientists need to meet and communicate with each other to develop trust that is so essential for the science to grow. It encourages meeting of different kinds of minds. Scientific meetings are a way to promote discussions on the frontier areas of science and technology. "In fact, the most stimulating scientific meetings that I have attended have taken the extreme opposite approach, intentionally mixing scientists with very different backgrounds and interests, convening them to produce a set of new ideas for attacking a challenging scientific puzzle," writes Bruce Alberts (7).

The design of scientific meetings constitutes an important part of science communication. 'Session-free afternoons' are important component of the scientific meetings; collaborations have germinated in many such afternoons. A place with minimum distraction is ideal for holding these meetings. Picturesque places do not necessarily distract, but can. Austere facility for holding the meets is really what one needs to encourage, but austere places are not very popular destinations. 'Large' meetings are often big distractions.

The US National Academy of Sciences organised a Colloquium a few years ago (8). Its aim was to improve the understanding between the scientific community and the public. Its objectives were as follows:

(1) Assess the scientific basis for effective communication about science,

(2) Strengthen ties among and between communication scientists,

(3) Promote greater integration of the disciplines and approaches pertaining to effective communication, and

(4) Foster an institutional commitment to evidence-based communication science.

The Colloquium discussed science related issues that focused on research in psychology, decision science, mass communication, risk communication, health communication, political science, and sociology. Questions such as — how science related decisions are taken, and what determines public attitude towards science —were raised in the Colloquium. One of the interpretations of the participants (both scientists and non-scientists) was that an individual's ideological view or cultural identity has greater influence on the opinions than an understanding of the fact.

Perhaps, scientists enjoy the positives that exist in life more than the others. They, perhaps, respond better to positive events and the opportunities around them. Perhaps, scientists wear rose tinted glasses better. According to some neuroscientists, "People with rose-tinted glasses are more responsive to positive things in the environment." Scientists love autonomy. Society has allowed them a great deal of autonomy. Society also expects them to understand that autonomy also means responsibility. Scientists need to understand that autonomy also includes accountability.

Science Needs Art

We need more than what is required for survival. Science needs art for its survival. There is positive synergy between art and science. Science gives perfection. Art brings into the fold of science the unexplainable world of magic. Science begins as philosophy, and ends as art. Art boosts attention, cognition, working memory and reading fluency. Art improves learning; it combines major tools (motor skills, perceptual representation, and language) that mind uses to acquire, store, and communicate.

Art, in association with science, can take us to the blind spots where we can't go alone. Niels Bohr, a lover of cubist paintings, once said, "We must be clear that when it comes to atoms, language can be used only as in poetry." For Bohr the invisible world of electrons was essentially a cubist world. He knew electrons could exist, as either particles or waves, but he also knew that the form they took depended on how we look at them, and their nature was a consequence of our observation.

Art is qualitative, and is generally associated with emotions. Science is quantitative, and is generally associated with intelligence. Art looks for questions, while science seeks answers. Science is more like a 'monologue'. Art is more like a 'dialogue'.

Both the artists as well as the scientists want to see what is beyond the obvious. Both require imagination. Both look for a chance to think outside the box. Both get distracted by the blind alleys of complex social problems. Both want to come out of these blind alleys in their own ways. Both draw inspiration from their conscious and unconscious realms.

Artists and scientists make a good natural partnership. Perhaps, practising science needs lot of knowledge and experimentation, while practising art needs lot of imagination and experience.

What about the incoherence, imprecision, abstraction and contradiction that art delivers? That is precisely the point. Isn't incoherence an essential aspect of the human mind? Don't we live in a world full of contradictions? The issue is how to make the 'two cultures' move forward? What kind of 'third culture' would close the gap between the 'two cultures'? Each side has something useful to offer to the other side. If they join hands, perhaps, one day we will find answers of some of our yet unresolved deep questions.

Leonardo da Vinci was a perfect example of the culmination of two cultures of art and science. da Vinci, an artist and a scientist, believed in connections. He believed that everything is related to everything else. He could see connections in things that were so different from each other. He related rocks with bones, soil with flesh, rivers with blood vessels. For him sight was the most revered sense. He thought, taking into consideration the strong connection between eye and the brain, visual information is "transmitted to the intellect via the receptor of impressions and the common sense, an area where all sensory inputs were coordinated."

Turbulent motion and vortex formation fascinated da Vinci. Based on his intricate understanding and observation, he drew images of water in spiral motion. His another fascination was force that gives life to all things. Bodily movements for him were motions of the mind. Due to intricate understanding of these motions he could capture continuity of motion in space. Da Vinci was proud of his artistic creation, as well as his ability to bend iron bars. Apart from painting, da Vinci studied biology, mathematics and engineering. His best tool was his mind.

Dilemma of Science

Since its inception, science is facing a dilemma, the cross between science and divinity. "Science without religion is lame, religion without science is blind", a popular aphorism by Albert Einstein governs the debate between science and divinity. Scientific and divine endeavour has been to explain the deepest puzzles of life, like our coming to this world, understanding the basis of life, mind-body relationship, among others. Both are in search of truth — one discovers it, while the other reveals it.

Initially people asked questions to the only gap-filler, God. Then science came to solve many of life's puzzles. It conveyed to the people that one doesn't need a divine designer to resolve all their problems. People have also understood that science doesn't have the answers of all their questions. As Freeman Dyson (9) says, we are standing in-between the two unpredictable. "God is what mind becomes when it has passed beyond the scale of our comprehension." Charles Darwin writes, "Ignorance more frequently begets confidence than does knowledge: it is those who know little, and not those who know much, who so positively assert that this or that problem will never be solved by science."

Many eminent practitioners of science are saying that there is no logical contradiction between science and divine, if both are clothed appropriately. Belief in the divine is not a scientific matter, and religious statement is not a scientific statement. "In science, a healthy scepticism is a professional necessity, whereas in religion, having belief without evidence is regarded as a virtue," writes Paul Davies (10). A majority of the people have no difficulty in accepting scientific knowledge and holding to religious faith.

William D. Phillips (11), a Nobel Laureate in physics, writes, "Why do I believe in God? As a physicist, I look at nature from a particular perspective. I see an orderly, beautiful universe in which nearly all physical phenomena can be understood from a few simple mathematical equations. I see a universe that, had it been constructed slightly differently, would never have given birth to stars and planets, let alone bacteria and people. And there is no good scientific reason for why the universe should not have been different." He then adds, "Many good scientists have concluded from these observations that an intelligent God must have chosen to create the universe with such beautiful, simple, and life-giving properties. Many other equally good scientists are nevertheless atheists. Both conclusions are positions of faith...I find these arguments suggestive and supportive of belief in God, but not conclusive. I believe in God because I can feel God's presence in my life, because I can see the evidence of God's goodness in the world, because I believe in Love and because I believe that God is Love."

It is absolutely fine if one is filled with doubt before accepting the other. Doubt is part of true faith. Religion risks turning into fundamentalism if it ignores scientific reasons. At the same time, science should try to listen what religion wants to say. Mutual incomprehension is not good for both. We expect science not to knock down the castles of human relations built by religion. "There seems an inherent human drive to believe in something transcendent, unfathomable and otherworldly, something beyond the reach or understanding of science." In other words, we may doubt the existence of God, but we don't mind believing in something, by whatever name we may call it. In this context, let me recall what John Naisbitt said, "The most exciting breakthroughs of the 21st century will not occur because of technology but because of an expanding concept of what it means to be human."

If God is unpredictable, so is science. Since God will not be dead, isn't it better if we change our conception about him? Can't we conceive god in a new perspective?

It is as difficult to prove God's existence, as his non-existence. Science's tools will never be enough to prove or disprove the existence of God. Belief in God is not a scientific matter, and religious statement is not a scientific statement. Conflict between science and divinity can be resolved amicably, if both understand the merits inherent in each side.

The real culprits are impure expectations from both the sides. It is not possible to prove everything. Let science confine itself to the realm of facts. Let divinity limit its reach to values. It is we who decide the confinements and limitations. "There are no 10 commandments in thermodynamics or molecular biology, no path to righteousness and charity and love in Euclidean geometry."

Impure expectations from both the sides clash, because both seem threatened by diversity each side offers. It is quite acceptable, if a scientist doesn't accept faith straight away, and vice-versa. If someone says, "earthquake is a divine chastisement sent by Gods for our sins", one may not support the argument of equating calamity with ethical failure. One may like to put a counter argument and say that "our sins and errors, however, enormous have not enough force to drag down the structure of creation to ruin." It is absolutely fine if one is filled with doubt before accepting the other. Doubt is part of true faith.

In the 19th century, Charles Darwin proposed a doctrine that for life to develop and evolve, God's intervention is not necessary. In the 21st century, some people argue that neuroscience could confront God, because it proposes that soul (or spirit) is not required to comprehend human traits such as love, morality and spirituality. They believe, these traits can be understood if one suitably understands the functioning of the brain. We may doubt the existence of God, but we don't mind believing in something, by whatever name we may call it.

The concept of body, mind and soul as a tripod has been perceived by many theists. It is believed that the science is a work of the mind, whereas the body is the physical resource used to transfer cosmic energy from one soul to another through a process called life. The theory of afterlife has been a matter of discussion of both the religionists and scientists.

Paul Davies believes that "both religion and science are founded on faith — namely, on belief in the existence of something outside the universe, like an unexplained God or an unexplained set of physical laws, maybe even a huge ensemble of unseen universes, too." This is perhaps the reason that a complete account of some phenomena can't be given independently by both science and religion. When Davies asked his physicist colleagues why the laws of physics are what they are, some said, it is not a scientific question. But one of the replies was, "There is no reason they are what they are — they just are." Davies concludes, "The idea that the laws exist reasonlessly is deeply anti-rational. After all, the very essence of a scientific explanation of some phenomenon is that the world is ordered logically and that there are reasons things are as they are."

Scientific Term That Ought To Be More Widely Known

EDGE (www.edge.org) asks – What scientific term or concept ought to be more widely known? Richard H Thaler responds thus: 'The Premortem'. We are so familiar with the term 'post mortem' that typically follows any disaster, along with the accompanying finger pointing. Such postmortem inevitably suffer from hindsight bias. "The handwriting may have been written on the wall all along. The question is: was the ink invisible?"

Gary Klein (12) came up with the term "The Premortem," which was later written about by Daniel Kahneman (13). The arguments is premortem helps to avert disasters. Two reasons are cited for this optimism:

(1) It can overcome the natural organizational tendencies toward groupthink and overconfidence. "After all, the entire point of the exercise is to think of reasons why the project failed. Who can be blamed for thinking of some unforeseen problem that would otherwise be overlooked in the excitement that usually accompanies any new venture?"

(2) It starts working with the premise that the project has failed. It then tries to find out what might have gone wrong, at least hypothetically. It is like assuming "the problem has been solved, and then ask, how did it happen?"

Thaler wonders, "And, how many wars might not have been started if someone had first asked: We lost. How?"

If postmortem is autopsy findings, then premortem is clinical diagnoses. Postmortem is performed after the death of the patient. Premortem is performed before the patient is autopsied.

Projects fail but we are reluctant to speak about the reasons they failed. We don't like to discuss our planning phase. We don't like to expose our weaknesses. We think exposing our weakness is bad even if it improves the chances of our success. Research conducted by Deborah J. Mitchell, of the Wharton School; Jay Russo, of Cornell; and Nancy Pennington, of the University of Colorado found that prospective hindsight increases the ability to correctly identify reasons for future outcomes. Premortem helps project teams identify risks at the outset, writes Gary Klein.

In a typical premortem, the project director tells others connected with the project that the project failed. He asks his colleagues to ponder over it, and tell him what might have gone wrong. He asks them to tell him also what they have not told before, for various reasons (impolite being one of them). All the reasons are then discussed in a group meeting. It is not easy to find fault in oneself, or in others, but in the larger interest of the project, one needs to come out of this mindset. Often small changes result into a major development. Premortem is a way to circumvent the need for a postmortem.

For Daniel Kahneman, premortem is an approach for making better decisions. It is form of time travel. It helps the decision makers to overcome blind spots. It helps people to optimise optimism. Even the pessimists get a chance to say things that otherwise would go to deaf ears. Rational pessimism often scores over irrational optimism. In many cases, the 'illusion of consensus' has to take the back seat. Premortem technique is a kind of cooperative competition'.

Science Refines Everyday Thinking

Science needs reasoned argument, constant scepticism, and open-mindedness. Science needs philosophy. "Philosophy can have an important and productive impact on science. Philosophy and science share the tools of logic, conceptual analysis, and rigorous argumentation. Yet philosophers can operate these tools with degrees of thoroughness, freedom, and theoretical abstraction that practicing researchers often cannot afford in their daily activities, " write Lucie Laplanea et al (14). But how in practice can we facilitate cooperation between researchers and philosophers, they ask. Taking a step towards the other, though seem easy, is not an easy task. They point out several obstacles.

Many philosophers think science has no relevance to their work; some are in favour of a dialog but are insufficiently trained to enter into dialog with the researchers. On the other hand, only a few researchers are aware of the work produced by philosophers on science.

To bridge the gap between science and philosophy, they recommend a few suggestions:

(1) Make room for philosopher's insight in scientific meetings, and also scientist's insight into philosophy conferences,

(2) Host philosophers in scientific labs and departments, and vice versa,

(3) Co-supervise PhD students,

(4) Create curricula balanced in science and philosophy,

(5) Journals where both science and philosophy contributions are discussed constitute an efficient way to integrate philosophy and science.

Laplanea et al quote Carl Woese to make their point: "a society that permits biology to become an engineering discipline, that allows science to slip into the role of changing the living world without trying to understand it, is a danger to itself."

Some people think, David Gelernter (15) one of them, that science has become an international bully. Gelernter says, to impress and intimidate scientists use their acquired power. He says they should be careful while using this power. "Too many have forgotten their obligation to approach with due respect the scholarly, artistic, religious, humanistic work that has always been mankind's main spiritual support." Due to this, Gelernter says, our subjective conscious experience is getting affected.

Science needs emotions. It is not a mere rational and objective enterprise. It needs 'head' as well as 'heart'. It needs to follow a different kind of religion. Einstein called it 'cosmic religion'. For Einstein, 'social and moral god' did not exist. His god did not reward or punish his creatures. For him, god was a 'conceivable mystery'. He thought that finer speculations in the realm of science can't be understood without a deep religious feeling. Writes Einstein in his book The World as I See It, "A knowledge of the existence of something we cannot penetrate, our perceptions of the profoundest reason and the most radiant beauty, which only in their most primitive forms are accessible to our minds — it is this knowledge and this emotion that constitute true religiosity; in this sense, and in this alone, I am a deeply religious man."

Thomas Kuhn (16), in his book The Structure of Scientific Revolutions, discusses about the progress of scientific knowledge. He is of the view that scientific progress is revolutionary, rather than steady and cumulative. He asserts that revolution occurs when a crisis is resolved by abandoning one paradigm and adopting another paradigm. Revolutionary science is unlike 'normal science'. In normal science scientists add to, elaborate on, and work with a central, accepted scientific theory; anomalies are resolved by a community of researchers who share a common intellectual framework either by incremental changes to the paradigm or by uncovering observational or experimental error. Kuhn believes that such 'paradigm shifts' are responsible for scientific progress, and our comprehension of science can never rely wholly upon 'objectivity' alone. Science must account for subjective perspectives as well, since all objective conclusions are ultimately founded upon the subjective conditioning/ worldview of its researchers and participants. "Our objective world is physical. Our subjectivity is a reflection of our inner landscape. Our subjective world can be examined and evaluated by us alone."

The rulebook of ethics includes many things. It includes honesty and integrity in acknowledging un-success and errors. Distorting or altering or omitting the facts that are not in conformity with applicable standards is unethical in all senses. It is the responsibility of the scientists and engineers to take care of the proprietary interests of others. And one of their responsibilities is to keep themselves updated about professional developments and practices.

Adequate familiarisation with societal demands is essential for a practical scientific literacy. The dilemma of science is that it can be used for good, as well as bad purposes. The same technique that promotes the advancement of medicine, can also facilitate production of biological weapons of mass destruction. Many of the methods for developing attenuated live vaccines against viral diseases can have offensive applications as well. Engineered biological agents could be worse than any disease known to man.

It is futile to imagine that access to dangerous pathogens and destructive biotechnologies can be physically restricted in a world where materials and technologies for biological weapons are easily accessible. It is also a fact that researchers can't stop doing research, because it can also do something we don't want it to do. Imagine for a moment that Louis Pasteur and Robert Koch were not allowed to do research to identify, isolate and culture disease-causing microbes to induce disease in a "native" host, because these techniques can also be used to produce biological weapons.

The only difference between a 'good' and a 'bad' approach is rationality in the design of experiments; one aims to develop better sanitation and hygiene practices, while the other wants intentional development of disease to be used as a weapon. The key issue is whether the risks associated with the misuse can be reduced, while still enabling critical research to go forward.

Ignorance, doubt, and uncertainty are important parts of scientific enquiry. Richard Feynman said, "Scientific knowledge is a body of statements of varying degrees of certainty– some most unsure, some nearly sure, none absolutely certain." We must recognize the lack of knowledge, and leave room for doubt in the pursuit of science. Feynman adds, "It is our responsibility as scientists, knowing the great progress and great value of a satisfactory philosophy of ignorance, the great progress that is the fruit of freedom of thought, to proclaim the value of this freedom, to teach how doubt is not to be feared but welcomed and discussed, and to demand this freedom as our duty to all coming generations."

One of the aims of science is to be able to predict events. Some events such as natural disasters are inevitable. In spite of the availability of best scientific tools and techniques, we fail to predict the unfortunate events. The scientific unpredictability could be due to some deficiency in our knowledge, or it could be due to inherent complexity in the system. In such situations, we depend upon probabilities. But the problem is our brains are not natively wired to assess risks and probabilities.

Mary Catherine Bateson (17) writes, "Our ability to avoid thinking about predictable events that have no predictable date is based on millennia of practice." Both interpretability and predictability depend upon the availability of data. But data have often failed us. "We just can't get enough data about our decidedly non-linear world to make accurate predictions." (18).

The synthesis between scientific spirit and humanism is necessary to refine our everyday thinking. Science has made enormous progress in the past century. Because of this development the relationship between science and society has changed substantially. In other words, science has changed the mind of our society. It has made the society more organic and has dispelled many traditional beliefs. It has also given us a wonderful vocabulary, and, as John Polanyi writes (19), "It is impossible to produce a vocabulary with which one can say only nice things." The responsibilities of a scientist, therefore, collectively as well as individually, need an update from time to time.

3

Engineering Designs the Vision of Life

Progress of any nation is driven by human curiosity and ingenuity. Engineering is one profession that has shown humanity the ways to meet its needs. It made the forces of nature work for the good of mankind. The biggest challenge for engineering profession has been its integration with the human needs. Engineering transforms technical conscience into applied realities. On the one hand, engineers are not limited by technology, and on the other, they are worried about the risks to the environment, health, sustenance and safety.

Engineer is a composite; "He is not a scientist, he is not a mathematician, he is not a sociologist or a writer; but he may use the knowledge and techniques of any or all of these disciplines in solving engineering problems." Engineering in the broadest sense, relates to the "development, acquisition and application of technical, scientific and mathematical knowledge about the understanding, design, development, invention, innovation and use of materials, machines, structures, systems and processes for specific purposes." Engineer works under various constraints: nature, cost, safety, environment, ergonomics, reliability, manufacturability and maintainability, among others.

This is how Neil Armstrong described himself, "I am, and ever will be, a white-socks, pocket-protector, nerdy engineer — born under the second law of thermodynamics, steeped in the steam tables, in love with free-body diagrams, transformed by Laplace, and propelled by compressible flow."

Some of the qualities we expect in an engineer are strong analytical skills, creativity, scientific insight, leadership abilities, high ethical standards, dynamism, flexibility, pursuit of lifelong learning, and dedication for public cause (1). It is a tall order, but then to create an ideal is always a tall order. Knowledge will never be complete, because perfection is unattainable and truth is unfathomable. Moreover, we do not want knowledge frontiers to advance so rapidly that we as a society lag behind.

The industry needs real-world engineers equipped to forge and deal with the complex interactions across many disciplines. Confident engineers are expected to foresee and manage unknown and unexpected problems. They are expected to appreciate, more than before, the human dimensions of emerging technology. They are expected to understand global issues and the nuances of working in a culturally diverse space. Industry needs "T-shaped" thinkers, deep in one field, but be able to work across all fields and communicate well. Industry needs engineers who take pride in designing a thing and manufacturing it. They need to appreciate that these jobs are also interesting as packaging. For the delivery of the goods packaging is not enough.

In the world of engineering, technical skills are not enough. Engineers are expected to bridge the gap between innovation and manufacturing. For being the 'best in the world' scientific discovery is important. It is also important to learn, as Paul Jacobs (2) says, how to work in interdisciplinary teams, how to iterate designs rapidly, how to manufacture sustainably, how to combine art and engineering, and how to address global markets.

People expect, from a low priced car, all the good things that make a car a good car. In fact, this kind of expectation tends to push up the bars of engineering excellence. Customer's priorities and demands are most important, and that decides what kind of trade-offs can be made to lower the costs. There is nothing like 'frugal' nation, as far as expectations are concerned. We all like low-cost-high-value products. Only frugal innovation that works on the principles of 'calculated trade-offs' succeeds. The question is - who needs the frugal innovation more – resource rich or resource constrained nation?

Engineers are expected to have the mindset of a polymath. Besides resolve and effort there is something else that makes a polymath. Ernst Schumacher writes, "Beethoven's musical abilities, even in deafness, were incomparably greater than mine, and the difference did not lie in the sense of hearing; it lay in mind." Leonardo Da Vinci, apart from painting studied anatomy, biology, mathematics and engineering. He was known for his intellectual, artistic and physical pursuits. If he was proud of his artistic creation; he was equally proud of his ability to bend iron bars.

Nurturing Future Engineers

The key traits of an engineer are integrity, grit and intelligence. For educators the challenge is in preparing engineering students for the "real world". The engineering curriculum is beginning to reflect the creative nature of engineering. More than before, the focus is on teaching and nurturing soft skills in students. At the undergraduate level, engineering schools are making efforts to ensure that the programme is interactive and engaging, with a clear focus on entrepreneurship and new technical skills.

There is emphasis on the changing role of teachers. Instead of the traditional role (lecturing to impart knowledge), the role of a teacher is now expected to be that of a coach and mentor.

Knowledge is available in plenty. Students want to develop their own individual learning programmes. For this, they expect guidance from their teachers. Good design is not enough, implementation is equally important. Operation is not sufficient, management skills are essential. Multi-disciplinary approach, including work environment, is important for nurturing future engineers. To learn new skills and competencies students may like to opt for more courses than is required to meet the requirements of the course. The mode of student evaluation need to broaden and change. The real world of engineering is no longer only expecting good grades. A big chunk of real world problems arise out of lack of knowledge of social engineering.

Industry feels engineering graduates lack soft skills, like communication and critical thinking. Industry wants decisive and insightful leaders capable of taking risks. An educated mind now needs wider competency. It includes thinking, reasoning and related skills. One needs to strengthen self-management skills, and that includes the ability to regulate one's behaviour and emotions. One must acquire the power of expressing oneself and impressing others. Understanding human relationships is as essential as understanding technology.

Engineering education is undergoing major transformation. The world is defined by increased connectivity, competition, and entrepreneurship. UNESCO predicts the future of education, as follows:

(1) Academic curricula will become more multi-disciplinary. This cross disciplinary learning and thinking can bring business and creative minds together.

(2) Education leaders will need to balance Massive Open Online Courses (MOOCs) and traditional learning.

(3) Student recruitment and retention will be more important than ever. Life experience prepares students to be healthy and dynamic, and that means sustainability and wellness will be the key components to college life.

(4) Higher education needs to invest in technology. The emergence of augmented reality devices will transform campuses. Institutions, therefore, will need to respond to the 'mobility shift', which will allow educators and students to be engaged from anywhere.

(5) Higher education will explore new funding models. Funding will be based on institution's responsibility to its students, and not on the basis of enrolment.

Andrea Bandelli (3) asks several distinguished academics - How they imagine higher education will look in 2030? A variety of opinions emerged. Some are:

(1) Intelligent machines will usurp the jobs of teachers. The concept of individual campuses will slowly disappear. The two-semester pattern will be replaced by year-round learning.

(2) The pedagogic pendulum will swing back towards the lecture, as the importance of an analytical mind becomes appreciated once more

(3) Exams that emphasise mastery of taught knowledge will no longer be the primary tool for judging student's performance.

(4) Technology has found a place in universities, but nothing significant has changed.

Kieren Egan (4) asks - Who is responsible for the educational ineffectiveness of our schools? Some opinions:

(1) Inadequately educated teachers,

(2) The absence of market incentives,

(3) The inequities of societies,

(4) The lack of local control over schools,

(5) The breakdown of the nuclear family and family values,

(6) A trivial curriculum filled only with the immediately relevant,

(7) A lack of commitment to excellence, mindless mass media.

One of the challenges of education is - How do we plan for something we can't predict? Another related question being asked is - Has education changed over the years?

Some of the findings of a study to evaluate the impact of the new outcomes-based criteria on engineering graduates from the US are summarized as follows (5):

1. Curricula have changed considerably with increased emphasis on professional skills and knowledge sets, like communication, teamwork, technical writing, and engineering design.
2. Teaching methods changed substantially; 'active learning approaches' such as group work, design projects, case studies, and application exercises were adopted.
3. Faculty are more involved in professional development activities.

4. There is reflection of the changes in the learning outcomes, particularly in social and global awareness, ethics and professionalism, group skills, and application of engineering skills.
5. Out-of-class experiences (such as internships, participation in design competitions, active participation in student chapters of professional societies) influenced student learning in important ways.
6. Employers' views differed; majority were satisfied with the ability of the students to learn, grow, and adapt, they observed improvement in team work and communication skills. Some employers, however, observed modest decline in problem solving ability.

In an editorial on reforming engineering education G Wayne Clough (6) writes, "Many also believe, I among them, that engineering graduates must also take on jobs outside of engineering, including jobs in non-profit and government policy areas where we desperately need people who can think clearly and logically and who understand technology."

Our problem is that most engineers are taking jobs outside the ambit of engineering; only a few want to work on the shop floor. The problem is not only with the graduates, but also the lack of availability of challenging jobs in the core engineering sector. One important suggestion Clough makes is that "we need to rethink which courses are really necessary and which ones can be reduced in scope or jettisoned to free up time for new material."

A makeover of engineering education is necessary, also believe WA Wulf and GMC Fisher (7), due to several reasons, including changing nature of international trade and the subsequent restructuring of industry. They see two interrelated problems: (i) fewer takers are opting for engineering education, and (ii) engineering institutions are forgetting the practice of engineering. Its consequences are: (i) large number of engineering colleges are closing down, and (ii) those coming out are ill-equipped to face the challenges of the industry.

The 'centre of gravity' of engineering education must shift, Wulf and Fisher point out. There was time when "the scientific and mathematical foundations of engineering" served the profession well. Now is the time for "empirical design methods based on experience and practice." Among the areas suggested for reform are faculty, curriculum, life-long learning, and diversity.

Many think, faculty-reward system does not value teaching and learning as an activity. They ask, "Can you imagine a medical school whose faculty members were prohibited from practicing medicine?" Another of their observation is that courses in life sciences should be the integral part of 'fundamentals'. They think, our creative field is deprived of a broad spectrum of life experiences that bear directly on good engineering design.

Timely advice of a mentor is essential part of engineering education, particularly at times when the students find the going tough, and are tempted to abandon engineering for an easier alternative. Engineering faculty must change their attitude, "and one good way to win hearts and minds is by their professional organizations–especially those positioned to reward individuals' achievements–conspicuously taking up the cause."

About the future of engineering education, Phil Kaminsky said a few things that deserve to be mentioned here (8). When he asked a Berkeley alumni – What has changed since the time she graduated (in 1989), she said that "the way we educate engineering leaders today has an entirely new look." The hallmarks of a good engineer are design thinking and entrepreneurial mindset. A good engineer has the capability to question 'status quo'. Writes Kaminsky, "we have found that the best entrepreneurs share a set of behaviours — they give and accept help, collaborate, communicate through stories, trust others, seek fairness, are resilient, have diverse personal networks, understand that 'good enough' is fine when time and resources are limited and believe that they can change the world."

"With the advent of the new technologies, individual centered education is only a matter of time," believes Howard Gardner (9). In an interview Gardner clarifies an important point, "I don't think you can do interdisciplinary work unless you've done disciplinary work." Without discipline one remains barbarian, he argues. He says, understanding is taking something that you've learned, a skill, a bit of knowledge, a concept, and applying it appropriately in a new situation.

What we do in schools is 'memorizing', and that has very little to do with 'understanding'. Gardner says, "The most interesting finding of cognitive science for education is that when we ask even the best students in the best schools to make use of the knowledge in a new situation, they don't typically know how to do it."

Gardner's way of understanding is (i) Any topic that's worth spending time on can be approached in many different ways, (ii) Provide powerful analogies or metaphors for what you're trying to understand, and (iii) Our understanding of a topic is rich to the extent that we have a number of different ways of representing it, and we can go pretty readily from one representation to the other.

The four things that are required before one enters college, according to Gardner, are:

(1) How to think scientifically, how to make sense of an experiment, what a hypothesis is,

(2) One needs to know something about the history of their country, something about the background, maybe a little about the rest of the world too,

(3) People need to know something about how to make sense of works of art because those are treasures of the culture, and

(4) One must have basic understanding of mathematics, one of the essential languages of science.

The curriculum needs greater emphasis on the amalgamation of art, technology, design and engineering and self-directed learning and creativity. "Humanities and Arts, Sciences, Engineering, and Medicine are branches from the same tree" is the theme of a report (10). It says, expectation from education is to empower the individual to separate truth from falsehood, bias from fact, and logic from illogic. Education has moved from 'integrative traditions' to 'disciplinary silos'. There is a need to return to the integrative model that seeks to bridge the integration of the Humanities and Arts with Sciences, Engineering, and Medicine in Higher Education.

Universities are becoming multi-varsities', "held together more by a unitary administrative structure and budget than by a collective commitment to truth or to a notion that knowledge is essentially integrated," wrote Clark Kerr more than 50 years ago.

Some of the rationale for integration are:

(1) It addresses the global challenges and opportunities,

(2) It prepares better graduates for employment and for engaged citizenry,

(3) It makes learning more engaging and relevant to students,

(4) It addresses the multi-dimensional challenges of our time - material, economic, environmental, social, cultural, technical, political, medical, aesthetic, and moral,

(5) It broadens interdisciplinary experiences to interact with strangers,

(6) It promotes diversity and inclusion,

(7) With better understanding about human history and culture, one can draw from a deeper pool of knowledge in understanding the context of their work and in solving problems, and

(8) It promotes innovative thinking that can lead to significant scientific breakthroughs; historical examples have shown that breakthroughs in science have been inspired by analogies provided by the arts.

"A well-educated person must have an understanding of the laws of physics, the structure of science, the ways of thinking that govern scientific and technological innovation, and the historical, social, economic, and political significance of science and technology in modern life. Artists and humanists must also understand the scientific and technological context of their craft, " is how one wishes to conceive education in the new world. It is important to recognize the critical limitations of particular ways of knowing, to achieve the social relations appropriate to an inclusive and democratic society, and to cultivate due humility. These are essential requirements for both personal and professional life.

Experiential Learning

For a long time, it was generally believed that the more a person knows, the better will be a person's life. The perception is now changing; the value of knowing things is decreasing, while the value of using knowledge is increasing. There is more premium on what someone can do rather than on what they know. Accordingly education is changing, with focus on experiential learning (11).

In the future education scenario, peers and mentors will have a bigger role. The focus will be on arousing intrinsic motivation. One may know things, but desirable will be who can do things and can generate ideas. To meet human dimensions of technology, we will require more than just engineers and technologists. And is rightly pointed out, "Engineering and science worry about feasibility, business and economics tackle viability, and the arts and humanities ask questions about desirability—all key aspects of true innovation." Soft science will strengthen the hands of hard science. The attributes of new syllabi will be entrepreneurial mindset, ethical behaviour, teamwork and leadership, global perspective, interdisciplinary thinking, creativity and design, empathy and social responsibility. More than the curricular details, important will be to make "engineering schools exciting, creative, adventurous, rigorous, demanding, and empowering."

Are we losing the ability to think about 'weird things', asks an expert. Thinking about 'Weird things', according to the expert, is to encourage one "to think about doing the impossible and to do the impossible in areas that matter to people who come from different backgrounds and geographic regions." Like someone with engineering background, having interest to cure cancer.

Katherine Banks (11) asks a very interesting question: why doctoral education focuses on turning out academicians when the majority of its Ph.D. students get jobs in industry or at the national laboratories? They get no training or opportunity for training in business, in communication, or in working in large teams. Should internship at the industry be a part of PhD programme?

It is easy to say such things, but so difficult to implement them in real situations. Take the case of an engineer-musician. One can be innovative, if the fundamentals of both engineering and art are sound. "If musicians are trained in the same way engineers are, they would first learn about the theory of vibration and sound and about note shapes and the natural frequencies of strings and columns of air. They then would learn music theory and orchestration, and finally, in their final year of school they would be given an instrument and taught to play scales. Musicians learn music by playing it, every day, and building up a repertoire." Says Richard Miller (11), "I think engineering is also a performing art in which you build up a repertoire in the same way. Musicians do learn theory along the way but not at the expense of playing music."

Merger of learning with experience is important. But it is not necessary for a student to have all of the fundamentals before they are allowed to explore applications, believes Miller. For Miller, getting rid of disciplinary departments is more important to creating a collaborative, innovative culture.

Education also requires new metrics for evaluating both student's and teacher's performance. This includes creating new categories that better reflect the overlap between building student success, building the institution, and producing nationally visible achievement.

Engineering Education

A meaningful mind of an engineer understands an individual's needs, experiences and emotions. Meaning is interactive, selective, and value-driven. "Meaning is like a large map or web, gradually filled in by the cooperative work of countless generations."

The future growth of our nation will continue to be driven by engineering. India recognizes that youth leadership has a crucial role to play as change agents for India's development. Young engineers have to adapt to the changing realities and keep pace with technological evolution. As nation builders, young engineers have to develop multi-faceted skills and competencies, going beyond their core specialization, including soft skill sets.

Birla Institute of Scientific Research (BISR), Jaipur conducted a pan-India quantitative survey covering 200 engineering students (in the second and third year of their undergraduate engineering courses). Some of the findings of the survey are:

(1) Majority of students feel the need of closer interaction and engagement with the industry; the form could be 'live industry project', change in curriculum in consultation with industry experts,

(2) Students like to discuss ideas with the faculty, preferably outside the classroom in a less formal atmosphere; students felt they have little scope of socialization outside the class room,

(3) A quarter of the students were unsure of what to do after graduation, and

(4) Majority of students decide to go for engineering when they were in High School.

Another survey conducted by Aspiring Minds (12) found that only 7 per cent engineering graduates are employable. It says, "Profit-hungry managements, lack of skill education, resplendent corruption, focus on rote-learning methods, and shortage of faculty (both in quantity and quality) are the major issues plaguing higher education."

The job profile of, and also the expectations from, an engineer are different from that of a manager. The skills required to become a good engineer and a good manager are different. The requirement of one job is 'focussing', whereas the need of the other job is 'overseeing'. As an engineer one is evaluated on the basis of performance. A manager is evaluated on the basis of group performance. For a manager, what matters most is relationship building and conflict resolving skills. Removing bureaucratic hurdles is one of the major responsibilities of a manager. One way is to learn the game of ethical politics to achieve what you want to achieve. Managers are required to be conversant with the changing norms of their playing turf. The job of a manager is like that of a caretaker. A caretaker takes care of what is in place and tries to make it more efficient. The problem is engineers find it difficult to play in all turfs, ethically or otherwise.

B. Michael Aucoin (13) says that transition from engineer to manager is possible. His recipe for successful transition includes the following fundamentals: mastering relationships; seeing the big picture; getting things done; communicating effectively; using assets wisely; and taking things to the next level. It means good engineers must also possess strong interpersonal skills, if they want to become good managers. Engineers are 'individualistic' by nature. They are required to develop the gelling capacity that is required in group activities, if they want to become good managers. Aucoin writes, "Engineers are uniquely qualified to be managers and leaders, in large part because they understand systems-thinking so well. Once you understand that organizations are simply systems of people, you've got it made."

What we need is innovative and socially responsible organisations and innovative leaders. Innovative organisations respect people. Carol Milano (14), based on a science careers survey, writes, "Innovative leader is the most powerful driver of the top companies." The top employers search for exceptional scientists who can bring fresh and original ideas to the company. A top scientist wants to be a part of the company that has the potential to make changes in the world. It is not enough for an innovative company to hire people who are only exceptional scientists. The person should also fit into the company's core values. The innovative company recognises the joys of uncertainty. Maintaining status quo doesn't satisfy an innovative company. In order to attract the right people, for example, a company advises the job seekers not to apply, if science is not their obsession, if one is content being the smartest person in the room, and if one is afraid to fail. In other words, if you are routine, you are not fit to work in an innovative company.

How about imitating the wonder that was India? Can't we regain that spirit and that confidence? Can't we, a country of 1.2 billion people, get back, literally, to Zero? India can literally 'get back to zero', provided it has young driving forces who can think 'outside the box'. What are the forces that drive the young to succeed in thinking outside the box? What must we do differently, as parents, teachers, mentors and employers to prepare young minds as innovators? Tony Wagner (15) says that teachers make the most critical difference in the lives of young innovators. These teachers are not ordinary teachers, but outliers, identified by the ways in which they teach.

The learning cultures that produce innovators are: collaboration, problem-based multidisciplinary approach to learning, taking risks and learning from mistakes, trial and error, creating real products for real audiences and encouraging intrinsic motivation rather than relying on extrinsic motivation (like "carrots and sticks, As and Fs"). "I found a kind of remarkable emphasis in the classrooms and among the parents of play, passion and purpose," writes Wagner. Whatever interested the young, they found teachers and parents supporting it. Young innovators want to make a difference more than they want to make money, observes Wagner. The young generation, he believes, can live on less. He, however, adds, "Those are easy things to say when you're in your 20s. When you're in your 30s — thinking about a family — that picture may change". He, however, feels today's generation is less materialistic. This 'connected generation' knows how to find support for what they want and need.

Wagner makes an important statement about Montessori schools: "Many of the most successful entrepreneurs and innovators today were, in fact, products of Montessori schools, where it is much more of a play-based form of learning." His message for the parents is that they should not try to overprotect their children. Helicopter parents hovering over their children all the time is certainly not an innovative idea. Mistakes strengthen self-confidence of the young innovators. For the young innovators, Wagner's advice is "to stay true to what your passion really is and your sense of a larger purpose in life".

We need to nurture future scientists. The Department of Science and Technology (DST), Government of India's INSPIRE (innovation in science pursuit for inspired research) programme is one such initiative. The purpose of INSPIRE is to inspire young minds to follow creative pursuits of science, and build the required critical human resource pool for strengthening and expanding the science and technology and R&D base of the country.

In our country we are producing, according to one estimate, around 9000 science, technology and engineering PhDs every year. It needs more. As per one projection, India hopes to produce up to 20,000 PhDs each year by 2020. More than the number, the quality of PhDs produced is important. One way of improving the quality is to revise the reward and recognition structure.

What the students think of doing PhD in India? In one survey, IIT students mentioned several reasons for not doing PhD in India. The reasons included too much time taken to complete PhD, too many pre-PhD courses, low market value, and uninspiring supervisors. Some supervisors, however, are too 'inspiring'; they don't hesitate to supervise more than a dozen students at a time. There is a need to change this trend. Let students also understand that completing PhD is their responsibility, not their supervisor's.

What the industry thinks of PhD holders? Many think, PhD holders are poor team workers. They are less adept at dealing with changing challenges. Many think some non-academic training is essential for PhD aspirants. They say, courses in marketing, communication and leadership are useful for a scientist alongside academic acumen of critical thinking and analysis. An issue that needs consideration is to "trample the boundary" among the scientific disciplines, because of the trans-disciplinary nature of science and technology.

MIT's state of the art report touches upon issues very pertinent to engineering education (16). It names a few institutions as 'current leaders' and 'emerging leaders' of engineering education. For case studies, the report selects four institutions; these institutions vary in terms of geographical locations, economic and institutional scale and approach.

One institute the report picked has design-centred focus; here the projects are integrated throughout the curriculum to reinforce learning. The institute does not offer conventional engineering disciplinary degrees; students study a common first year and then specialize within one of four multidisciplinary 'pillars' – such as Engineering Systems & Design – for the remainder of their studies. The government has also offered the institute the autonomy and flexibility needed to establish an educational culture and approach that is both unique and world-class. The report also considers an engineering education system that operated in very traditional mode until a few years ago. Then they adopted a 'radically different approach' that entails a student "to work effectively with people from a range of backgrounds and perspectives." Its Integrated Engineering Programme has two major components: (1) a common curricular structure applied across all engineering departments during the first two years; a series of five-week curricular cycles where students would spend four weeks building critical engineering skills and knowledge that they would subsequently apply in a one-week intensive design project, and (2) multidisciplinary experiences to work effectively with people from a range of backgrounds and perspectives.

The programme of yet another technical university, known for its ethos of openness for creativity to flourish, is designed and delivered, keeping in view (1) deep disciplinary knowledge, (2) design-centred learning, (3) significant opportunities to apply their knowledge to real engineering problems, and (4) online learning. It has decentralized environment; "a collection of islands and small kingdoms," where "the bright spots are scattered" and pedagogical practice varies considerably across the institution.

The case studies brought out a few common features that facilitate best practice in engineering education: a collegial and exploratory educational culture, student engagement in and understanding of new educational approaches, and in-house development of new tools and resources to support and advance the educational approach.

The report lists key challenges engineering education will face in the coming decades. These are:

(1) The alignment between governments and universities in their priorities and vision for engineering education, like purpose of engineering undergraduate education, government regulation and national accreditation, unpredictable nature of higher education funding, and commodification;

(2) The challenge of delivering high-quality, student-centred education to large and diverse student cohorts;

(3) The siloed nature of many engineering schools and universities that inhibits collaboration and cross-disciplinary learning;

(4) Faculty appointment, promotion does not appropriately prioritize and reward teaching excellence.

(5) The quality of education can't be measured merely on the basis of staff-to-student ratios, and graduate employment profiles.

(6) "The scholarly work going on in engineering education is not translated back into the lecture room, it's always theoretical."

(7) The discipline/department-based structure of many engineering schools and universities are holding back innovation and excellence in engineering education.

According to an expert, "impenetrable departmental silos that make it almost impossible to offer students real choices that are not completely disconnected from the rest of their curriculum." The engineering institutions that are established in recent years have carefully excluded disciplines that are no longer relevant for resolving many of the current problems.

Most engineering programmes are research-led and are designed such that faculty gets adequate time for research. It recognizes the fact that becoming an 'adequate' teacher and researcher is not easy. Evaluating an 'adequate' researcher is comparatively easy than evaluating an adequate teacher. The report observes: "measuring the impact we have on our students, how much they are actually learning, is something that we as a community do very badly." The report further notes: "what determines reputation of a university continues to be research. Unless this changes, it is difficult to see how there will be an extensive change to teaching."

The report says that curriculum will continue to feature areas such as user-centered design, technology-driven entrepreneurship, active project-based learning and a focus on rigor in the engineering fundamentals. In addition, the following curricular themes will get increasing prominence:

(1) Student choice and flexibility; students will be allowed to choose a pathway that suits their talents and interests,

(2) Multidisciplinary learning; both within and engineering,

(3) The role, responsibilities and ethics of engineers in society; a greater focus on solving human challenges and the problems facing society would be the hallmarks of good engineering programmes,

(4) Global outlook and experiences; providing students with a range of opportunities to work across nationalities and cultures,

(5) Breadth of student experience; increasing emphasis on experiences outside the classroom, such as work-based learning.

There are many questions we must ask ourselves, for the meaningful nurturing of future minds:

- Are the minds of engineers a combination of opportunity and resources?
- Are we cultivating the right kind of engineering mindset?
- Do only the academic grades reflect the quality of an engineer?
- What is more important for an engineer – insight or precision?
- Can a person trained to solve expected problems deal with unexpected problems?
- What additional efforts are required to impart practice- based experiential knowledge?
- What a general engineering toolkit must contain?
- Is there a need for various specialized engineering streams at the undergraduate level?
- Shouldn't the practice of industry mentoring be taken more seriously?
- Why most engineers don't take as much pride in designing a thing and manufacturing it, as they take pride in packaging it?

Research and Innovation

"The most important instrument in research must always be the mind of man," wrote W. I. B. Beveridge years ago (17). This 'instrument' is still most sought-after in research and innovation. Mind is the most prized asset of man. Knowledge is being generated at a rapid rate, and we are asking ourselves, can knowledge ever exceed man's head? If man's head expands more than its actual utilisation, will there be problems?

It is important to know what is already known and what is yet to be known. Knowing the known, as well as the unknown, requires curiosity, attentiveness, patience, and scepticism. A knowledge seeker is never sure if he has arrived at the right conclusion, but he continues to work on the premise that one day he will arrive at the right conclusion.

There are known unknowns. There are also unknown unknowns. Cognitive scientist Gary Marcus (18) says, we should be more worried about the 'unknown unknowns', as 'unknowns' are associated with gains and risks. Marcus says that unclear risks that are in the distant future are the ones we take less seriously. We have the tendency to discount the future impact of the risks, because we don't know them. We like to believe that we live in a just world, where there is little scope for unclear risks. Often, unclear risks cause more serious problems than clear risks. We often tend to ignore the future risks. We are forced to remember them only after the 'risks' happen. Writes Marcus, "What we really should be worried about is that we are not quite doing enough to prepare for the unknown."

Creative minds are wired for rapid and fluid thinking. These minds can see things in new ways and have the ability to make quick associations. A creative person can connect the 'seemingly dissociated', and can see 'patterns' where others see chaos. According to psychologists, creativity requires intellect to generate unusual associations and analogies, as well as rich imagery.

Like the minds of creative people, we all get flooded with ideas. But creative minds have the ability to control the flight of ideas into a meaningful trail that many of us don't have. Creative minds are vulnerable and sensitive, like other normal minds are. They also suffer from high and lows and mood swings. They also lack confidence and need encouragement to overcome their doubts. But creative people know how to navigate the tides. Many instances indicate that creative people see the world in unconventional ways, and are often at odds with the world and themselves.

Alex Osborn observed more than six decades ago, "It is easier to tone down a wild idea than to think up a new one." How effective are Osborn's concept in the present environment? Take a look! When people want to extract good ideas from a group, they still obey Osborn's cardinal rule, censoring criticism and encouraging the most freewheeling associations. Several recent studies, however, have exemplified 'sobering refutation of Osborn'. Studies have indicated that brainstorming makes individuals less creative. They say, fewer ideas are generated when people pool their ideas, compared with when they work alone. But then there are many who believe creativity is a group activity.

Many ideas emerge, more by happy accidents than by design or deliberation. Some discoveries are not unintended, but the discoverers managed to do so via some unexpected route, and usually not without considerable trial and error. This is not to say that these discoverers are lucky scientists. Creativity requires special abilities or traits that sets a creator apart from others.

R Keith Sawyer (19) unfolds some hidden secrets of creative minds. He says that there is nothing like a "full-blown moment of inspiration." Ideas don't magically appear in a genius' head from nowhere; they always build on what came before. An idea may seem sudden, but in reality our minds have actually been working on it for a long time. Ideas emerge from a chain reaction of many tiny sparks. Often we fail to gauge which spark is the brightest. One of the advices of Sawyer is to develop a network of colleagues and schedule time for freewheeling and unstructured discussions. Creativity, like many other pursuits, is hard work. "Look at what others in your field are doing. Brainstorm with people in different fields. Research and anecdotal evidence suggest that distant analogies lead to new ideas," writes Sawyer.

Dean Simonton (20) argues that scientific creativity is a joint product of logic, chance, genius, and zeitgeist – 'with chance the first among equals'. A genius is a 'creative elite'. The zeitgeist ('the spirit of the times') perspective suggests that the inevitable product of scientific creativity arise from emerging social needs. This perspective reduces a genius to a 'mere agent of sociocultural determinism'. Some say, all ideas are already in the air, and for someone to pick them doesn't require great intelligence. Some cite instances of 'multiples' (for example, theory of evolution by natural selection by Darwin and Wallace almost at the same time) to support their argument.

When you focus your attention on something, you only see a very small fraction of your field of vision, because your brain gets filled in with everything what you think is there. In doing so you may miss the disruptive things that are happening at the periphery. Says Joichi Ito (21) "You've got to be antidisciplinary, because if you're in a discipline and you're worried about peer review and you're knowing more and more about less and less, that's by definition an incremental thing." Ito says, when you are antidisciplinary you have the freedom to connect things together that aren't traditionally connected. Ito promotes 'build and then think about the business model' philosophy. If a thing works, its theory can be sorted out later. Ito believes in 'demo or die', rather than 'publish or perish'. Some of Ito's prescriptions are: stop focusing on individuals and start focusing on communities; stop focusing on top-down and focus more on bottom-up; stop focusing on single experts and start focusing on the cloud.

"An antidisciplinary project is when you can't tell who the designer is and who the engineer is, and the engineer knows designing, and the person who's dancing is going to be the one also doing the molecular biology".

The other point Ito conveys is that you may not have a first class degree, but you know what really you are good at, and what you are obsessed with. What Ito wants to convey is that to steer the innovation boat you also need few 'misfits'.

Innovation is not for everybody. You can't just give somebody a bunch of free time and then expect him to come up with a brilliant idea. There is need for togetherness, but togetherness of a different kind.

The 'magic moment of innovation' comes when one frees oneself from structured thinking. Such thinking is divergent, and is open to alternate perspcctives. It identifies goals, examines assumptions, accomplishes actions, and assesses conclusions. The neural pathways of such thinkers operate in unconventional networks. Such thinking does not come naturally; the harder you try to think differently, the more rigid the categories become.

An emerging element of innovation is openness. Open innovation organisations don't confine themselves to inner means for solving problems. They go out, wherever they have to, to find solutions. Once they know where the solutions are available, it doesn't matter if they are available with a competitor; they make all efforts to obtain it, not necessarily by adopting buying strategies.

The challenge before the innovator is to anticipate or evaluate the right status of a discovery. The innovator wants to understand the value of incentives, for furthering her/his purpose. She/He needs to respect the essence of time and speed. She/He needs to recognise the opportunities, as well as the risks, involved with the innovation.

Is creativity the preserve of a few geniuses or can be found, in some form, in all individuals? Rex Jung (22) studied the interaction between creativity and intelligence, mainly among college students. Jung says, "Creativity and intelligence are linked at lower levels of IQ, but above a certain threshold, they don't necessarily go hand in hand." He claims that people with high and low IQs appear to use their brains differently to achieve creativity. Among the ways suggested to foster creativity, besides purpose and intention, are building motivation, developing self-management, and leading a fulfilling life.

Howard Gardner thinks that talent and expertise are necessary, but not sufficient to make someone original and creative; "achievement is not just hard work: the differences between performance at time 1 and successive performances at times 2, 3, and 4 are vast, not simply the result of additional sweat." Gardner believes that the answer to the question — why some minds are more beautiful — will come through a combination of several findings. Genetics will give some insight on why highly talented individuals have a distinctive, recognisable genetic profile. Neuroscience will explain why there are differences in structural or functional neural signatures. Cognitive psychologists will tell us more about the psychology of motivation of talented individuals. They will tell us why the talented individuals develop passion to master their art.

Knowledge is not for pleasure and vanity only, believed Francis Bacon. Curiosity is hunger for exploration. Hunger for exploration sometimes is based upon natural curiosity and inquisitive appetite, sometimes to entertain the mind, sometimes for ornament and reputation, most times for lucre, and seldom to give a true account of the gift of reason, to the benefit and use of men, believed Francis Bacon. Bacon's advice was to separate and reject vain speculations and whatsoever is empty and void, and to preserve and augment whatsoever is solid and fruitful.

Bertrand Russell writes, between science (definite knowledge) and theology (what surpasses definite knowledge) there lies 'No Man's Land'. And this 'land' belongs to philosophy. One needs to visit this land for innovation nourishment. This land is quite useful for achieving excellence. It gives the ability to see things from various perspectives and with an open mind.

The Fourth Industrial Revolution

The first industrial revolution was mainly concerned with the mechanization through water and steam power. The use of steam-powered engines and water as a source of power helped agriculture greatly. Textile industry was one of the major beneficiaries of this revolution. The defining characteristics of the second revolution was mass production and assembly lines using electricity. The third revolution, often referred to as the Digital Revolution, relates to the adoption of computers and automation. The fourth industrial revolution, Industry 4.0, takes automation to a new level with smart systems fuelled by data and machine learning. This manufacturing revolution has the potential to produce a macroeconomic shift, boost employment, productivity, and growth.

Industry 4.0 brings the era of digitization of manufacturing. The 'smart factory', a combination of cyber-physical systems, the Internet of Things and the Internet of Systems, is an attempt to make the process of manufacturing more efficient and less wasteful. Few of its possible applications include (23):

(1) Identify opportunities for manufacturers to optimize their operation by knowing what needs attention.

(2) Optimize logistics and supply chains to adjust and accommodate when new information is presented.

(3) Autonomous equipment and vehicles to streamline operations.

(4) Affordable robots to support manufacturers.

(5) Additive manufacturing (3D printing), from prototype development to actual production.

(6) Internet of Things and the cloud.

Technology trends transforming industrial production (Boston Consulting Group https://www.bcg.com/en-in/capabilities/operations/embracing-industry-4.0-rediscovering-growth.aspx) include:

Big Data and Analytics - collection and evaluation of data from many different sources to support real-time decision making.

Autonomous Robots – to interact with one another and with humans for a range of capabilities used in manufacturing.

Simulation - to leverage real-time data and mirror the physical world in a virtual model for increasing quality.

Horizontal and vertical system integration - cohesive, cross-company, universal data-integration networks to evolve truly automated value chains.

Industrial Internet of Things - decentralized analytics and decision making, enabling real-time responses.

Cybersecurity – to secure reliable communications as well as sophisticated identity and access management of machines and users.

The Cloud – for increased data sharing across sites and company boundaries enabling more data-driven services for production systems.

Additive Manufacturing – to produce prototype, individual components, small batches of customized products.

Augmented Reality – to support a variety of services, such as selecting parts in a warehouse and sending repair instructions over mobile devices, to provide workers with real-time information to improve decision making and work procedures.

Industry 4.0 will bring changes. "The coming of steam power and the rise of robotics resulted in the outright replacement of 80 to 90 percent of industrial equipment. In coming years, we don't expect anything like that kind of capital investment. Still, the executives surveyed estimate that 40 to 50 percent of today's machines will need upgrading or replacement," write Cornelius Baur and Dominik Wee (24). The new manufacturing business models recognize the role of new competitive challenges. "Today, many manufacturing companies have deep expertise in their products and processes, but lack the expertise to generate value from their data." The need to monetize the value of expertise is yet another challenge some companies face. "Hunt for the best digital talent" is another formidable challenge in a fiercely competitive world of manufacturing.

The potential benefits, as enumerated by Martin (25) include the following: (1) self-optimization will lead to almost zero down time in production, (2) flexible customer-oriented market will reduce the gap between the manufacturer and the customer, (3) because of the new requirements and skills education and training will take a new shape. The challenges envisaged are: (1) IT security breaches and data leaks, cyber theft, (2) highly capital intensive, (3) threat to customer's privacy.

How Industry 4.0 will impact future work force is a major issue of discussion, as employment related issues are not yet very clear. The general consensus is that machines should not take over the industry. The following are some of the observations regarding the demographics of employment:

(1) Big-Data-Driven Quality Control will decrease the need for quality control workers, whereas the demand for big data scientists will increase.

(2) "Robot coordinators" of the smart devises will replace some production workers (such as in packaging).

(3) Self-Driving Logistics Vehicles with the assistance of big data will make many drivers obsolete.

(4) Need for industrial engineers to simulate productions lines will increase.

(5) Smart devices will make prediction of failures easier. These being self-maintaining types, the number of traditional maintenance technicians will drop.

Is India prepared for the Change?

One may ask – should India adopt Industry 4.0? This question is no longer relevant. The question one must ask is – Is India preparing for the change? The answer is – yes, India is preparing for the change. Government of India plans to increase the contribution of manufacturing sector to 25% of Gross Domestic Product (GDP) by 2025, from the current level of 16%. India is undertaking the Make in India programme. One of its prime agenda for development is Smart Manufacturing. It is setting up facilities to help SME's. Internet of Things hubs are being set up in some states with the active participation of private sector. FMCG, Telecom, Healthcare are some of the sectors that have adopted Industry 4.0. (26).

The Fourth Industrial Revolution faces several moral dilemmas (27). Dalmia and Sharma believe that humanity will be on the cusp of re-thinking morals – as Ethics 2.0. They put forward some very interesting questions related to moral dilemmas vis-à-vis Industry 4.0. Should gene editing be legal to manipulate the human race and 'create' designer babies? How do we iron out biases that creep in man-made AI systems? What if our Kindles were embedded with facial recognition software and biometric sensors, so that the device could tell how every sentence influenced our heart rate and blood pressure? What about the rights of humans to marry robots and of robots to own property? Should a highly advanced Cyborg be allowed to run for political office?

An open letter signed by several stalwarts highlight the benefits of AI. They, however, are wary about its potential pitfalls. " The potential benefits are huge, since everything that civilization has to offer is a product of human intelligence; we cannot predict what we might achieve when this intelligence is magnified by the tools AI may provide, but the eradication of disease and poverty are not unfathomable. Because of the great potential of AI, it is important to research how to reap its benefits while avoiding potential pitfalls."

Many ethical questions can be amicably resolved if we understand that the fourth industrial revolution is about empowering people, not the rise of machines. Machines can never supply the judgment and ingenuity, people can.

Frugal Engineering

Frugal engineering is "the art of doing more (and better) with less." It is making the best use of both time and money. It is making the best use of what you have. 'Bigger is not always better' for the resource constrained environments was the message Carlos Ghosn, the then CEO of Renault-Nissan gave to the world, and coined the term 'frugal engineering'.

"Frugal innovation is not just about doing more with less. It's about learning how to innovate under severe constraints and turn extreme adversity into an opportunity for growth. But it's hard for Western executives to cultivate this frugal, flexible and inclusive mindset — which we call jugaad — in resource-rich and relatively stable Western economies." (28).

Nirmalya Kumar, Phanish Puranam, Phanish Puranam (29), based on extensive research on Indian companies say that India has good potential and capability for frugal engineering. The purpose of frugal engineering is "simplicity not sophistication). However, the challenge is India "demands everything in the world, but cheaper and smaller." Their recipe for success in this endeavour stands on six principles:

(1) Robustness (harsh environment, huge variances in operating conditions affects the priorities of product development and innovation),

(2) Portability (space constraints and poor transportation links),

(3) Defeaturing (developing and adapting products for the local market avoiding implementing features that do little to enhance the product),

(4) Leapfrog technology (adopting technologies that are relevant for India and not necessarily to several other countries, like battery-powered, ultra-low-cost refrigerator resistant to power cut),

(5) Megascale production (lower costs compared to prevailing world prices due to megaproduction),

(6) Service Ecosystems (low cost but with broadened product appeal and also financing options).

One of the purposes of engineering is to serve the people who are drowned in poverty, and also those who are moving out of poverty. The so-called demographic 'bottom of the pyramid' has developed a good sense of expectation. In spite of their limited purchasing power, they are demanding customers. It means that low-cost engineering is not enough anymore. The challenge is to earn high volume profit in lower price market. The challenge is to minimize non-essential costs, while maximizing the value the customer gets. An innovation of this kind needs top-down support.

A consumer wants the best. Products lacking in safety and quality are bound to get rejected. One of the hallmarks of good engineering design is that it should be affordable. People expect from a low priced car all the good things that make a car a good car. In fact, this kind of expectation tends to push up the bars of engineering excellence. Customer's priorities and demands are most important, and that decides what kind of trade-offs can be made to lower the costs.

There is nothing like 'frugal' nation, as far as expectations are concerned. All like to have low-cost-high-value products. Only frugal innovations that work on the principles of 'calculated trade-offs' succeed.

Technology Hubs and Start-ups

Tech hubs and start-ups are created to help ideas germinate and companies prosper. It needs an environment specifically targeted at helping young technology companies thrive by encouraging experimentation, and helping firms network with other like-minded individuals and enterprises.

The time scales of innovation are city dependent. As population increases, it becomes more connected. As a result of this, the time scale of innovation in big cities becomes shorter. It has also been suggested that the pace of urban life in cities above a certain population increases, as the size of the city increases. The growth and sustainability of the cities, however, are constrained by the availability and mobility of resources and their rates of consumption. The system collapses when it runs out of resources; supply lines are choked and fail to meet the increasing rates of consumption. Innovation as well as knowledge creation and their application at problem solving are viable means to avoid this crisis.

Innovation has the potential to bring major qualitative changes in the life of the populace as well as infrastructure. "As we approach the collapse, a major innovation takes place and we start all over again", explain Geoffrey West and Luis Bettencourt (30). Big cities provide better opportunities to creative people for innovation and wealth creation. But the paradox is that as the size of the city increases, along with it increases the number of criminals, sick, job-seekers, and the amount of waste, etc. West and Bettencourt argue that up to 85 per cent of the character of a city is determined simply by its size. It means that only 15 per cent of a city's character is distinctive. It means that if a city doubles in size, it needs only 85 per cent more infrastructure. It also means that crime and disease increase by 15 per cent above the doubling rate. This paradox prompts us to ask another important question: What sized cities are most appropriate to set up innovation centres?

In the recent times there has been a massive migration of rural population to urban centres. More than half of the world's population now lives in cities. One of its implications is that towns are becoming cities, and cities are getting transformed into metropolis. As cities grow, its character changes, in terms of its consumption pattern and population behaviour. They tend to attract creative people, its potential for innovation increases, new wealth and resources are created, and at the same time, crime, pollution, and disease increase. A city starts to lose its identity when it becomes a megacity. As the city grows, it optimises the delivery of the social services. This is due to the economies of scale in infrastructure and facilities. The change in the character of a city is primarily driven by innovation, and is followed by wealth creation. The city is expected to grow continuously, and for that it needs an "accelerating treadmill of dynamical cycles of innovation."

Innovation city requires a strong education system to provide suitably skilled graduates, cutting-edge digital connectivity, and good infrastructure. Developed countries like Switzerland, Finland, Israel, and USA rank high in the list of innovation countries as they invest more heavily in human capital and research, have stronger institutions and markets, and have greater outputs in terms of knowledge, technology, and creativity. Among the top Tech Hubs around the world are Beijing, Los Angeles, Berlin, Bengaluru, Tel Aviv, Boston, London, Moscow, Lisbon, New York, Singapore, Amsterdam, Prague. India is home of the youngest entrepreneurs in the world and is one the fastest growing Tech Hub in the World.

Incubators are one of the prominent tools to promote S&T-driven entrepreneurship in India. These are designed to nurture start-ups by providing infrastructure and support services and the linkages with other participants of the innovation network. The report entitled Enhancing S&T- based entrepreneurship: The role of incubators and public policy by Kavita Surana, Anuraag Singh and Ambuj D Sagar highlights the evolution of the country's innovation ecosystem. The Department of Science and Technology, Government of India has established a few Centres for Policy Research at various academic institutions. This report was prepared by CPR's IIT Delhi centre. The report broadly explores the current incubator landscape, critically examines the strengths, weaknesses and barriers to success. The incubators are located around Hyderabad, Chennai, Bengaluru, Delhi, and Ahmedabad. Information Technology and Biotechnology are the two dominating sectors. The report provides a few recommendations and guidelines for the success of Startups.

1. Students lack skills in S&T driven entrepreneurship and innovation. It suggests development of new courses to fill this gap. This includes inter-disciplinary courses, considering the needs of the society and understanding of the market.

2. Faculty have limited knowledge of industry and have few incentives to launch startups. Incentivize faculty engagement in innovation and entrepreneurship, like flexible hiring and promotion policies, revising rules for 'study leave', support and fund sabbaticals for launching startups, etc.

3. Incubator activities are misaligned with poorly-defined incubator goals. Some of the measures suggested are inclusion in the goals sectoral and geographical development, co-location with academia, startup stage (i.e., idea stage, early stage, or growth stage), and offering periodic guidance to incubators in articulating and adapting goals and activities to the changing context of innovation and market needs.

4. Incubates lack skills in entrepreneurship. Build community, strengthen networks with industry, investors, and provide mentoring in a more targeted manner, appraising startup business models, writing business proposals, applying for grants, developing communication skills, etc.

5. Weak incubator management. Involve experienced professionals with business, market, and S&T experience in incubation management, developing strong skills in leadership and marketing, expanding well-performing incubators and strengthening their linkages with other incubators, phasing out funding to incubators that fail, initiating and incentivizing public-private models for incubator financing, and passion for S&T-innovation.

6. Startups/Incubatees lack support service to validate S&T-based ideas. Help the startups in formulating mechanisms for testing and validation of new technologies, establish technical services, legal and patenting services, market research, etc, facilitate connections between publicly-funded startups and public sector to secure advanced market commitments including provisions for public procurement of technologies, leverage alumni networks to create venture funds for supporting startups in areas of expertise of the university or of the incubator.

The primary objective of 'Make in India' initiative is to make India a global manufacturing hub and increase its global trade significantly in the near future (www.makeinindia.com). It intends to develop the manufacturing sector, foster innovation, skill development, implement intellectual property rights, and promote foreign direct investment. Its purpose is change of image of India from a 'regulator' to a 'facilitator'.

India can do it. It has enough market opportunity and promising local talent. It has low-cost manufacturing hub and untapped economic opportunity. Mudambi et al (31) have identified three key value-chain practices for achieving success in India:

(1) Collaborate horizontally with Indian network orchestrators to achieve localization advantages,

(2) Partner vertically with local suppliers to achieve local and global sourcing advantages,

(3) Leverage local and global products simultaneously. Foreign multinationals should first introduce their latest products in India for the upper-income segment and then quickly localize to accelerate market penetration across multiple income segments.

National Academy of Engineering

National Academies play a significantly important role in the development of a nation. They can be one of the strongest arm of a nation. Bruce Alberts (32) writes, "Critical to success (of the Academy) is its unique mission to provide independent, evidence-based scientific advice to the nation's policy-makers." Its elected members set a standard for scientific excellence that has important implications for the future. He further writes, "An academy's advisory roles are crucial because science's remarkable understandings about how the world works have profound implications for policy-makers."

The twentieth century represents a century of innovations. It was an era of engineering triumph with profound and widespread effect on society. The innovations transformed the way people live. The list, as drawn by US National Academy of Engineering, includes: Electrification, Automobile, Airplane, Water supply and distribution, Electronics, Radio and television, Agricultural mechanization, Computers, Telephone, Air conditioning/refrigeration, Interstate highways, Space flight, Internet, Imaging, Household appliances, Health technologies, Petrochemical technology, Laser and fibre optics, Nuclear technologies, High-performance materials. It was an era of 'affordable technologies'. (33).

Vision of engineering in the new century, a report prepared by the US National Academy of Engineering (34) envisages myriad challenges for the engineers of 2020. "Engineers will be expected to anticipate and prepare for potential catastrophes such as biological terrorism; water and food contamination; infrastructure damage to roads, bridges, buildings, and the electricity grid; and communications breakdown in the Internet, telephony, radio, and television. Engineers will be asked to create solutions that minimize the risk of complete failure and at the same time prepare backup solutions that enable rapid recovery, reconstruction, and deployment." Social issues - intellectual property, multilingual influences and cultural diversity, moral/religious repercussions, global/international impacts, national security, and cost-benefit constraints – will continue to drive engineering practice. Multidisciplinary teams will be needed to pursue international collaborations. The team members are expected to understand the 'language' of such a multidisciplinary team. 'Language' in this context includes communication skills in a techno-bulged world, an understanding of the complexities associated with a global market and social context, and flexibility and mutual respect among the team members. In this world the emphasis will shift from "low-wage, mass-production manufacturing jobs" to "creating a workforce and business environment that prospers in a mass-production-less economy."

In a dynamic environment of the future "engineers will need dynamism, agility, resilience, and flexibility." The future engineers are expected to learn continuously throughout his or her career, not just about engineering, but also about history, politics, business, and so forth.

The Indian National Academy of Engineering (www.inae.in), the premier Engineering Academy of India comprises most distinguished engineers and technologists. It functions, as an apex body, to promote the practice of engineering and technology for solving the problems of national importance. The Academy provides a forum for futuristic planning for country's development requiring engineering and technological inputs, and bring together specialists from such fields as may be necessary for comprehensive solutions to the needs of the country.

The INAE has prepared a vision document (35) to highlight societal and engineering challenges. Foremost among the engineering challenges pertinent to India are: providing quality education, innovation-centred research, aspiration-driven manufacturing, and value engineered public services. The growth of socio-economically vital sectors have to be driven through efficient mobility and connectivity through multimodal road and high speed railway networks, well connected airways, low carbon energy automobiles, and access to space and deep-sea environments. The INAE foresees the need for more robust and technology-backed disaster warning systems, supported by fail-proof mitigation measures for ensuring secured lives for its people. The basic tenets for sustainable living, identified in the Vision Document 2037 are: food and agriculture, poverty alleviation, habitat, water & sanitation, energy, and healthcare.

Food and Agriculture: Nutrition at affordable cost needs to be met in a scientific manner, keeping in mind the food habits in different parts of the country. It will require a different kind of 'green revolution' and engineering intervention. Since arable land is scarce, reclaiming fallow or saline lands for agriculture purposes will be needed on a much bigger scale. Major issues will be enhancing crop productivity and efficient use of water and fertilizer. Efficient market access of agricultural products, problems of perishability and storage will be the additional challenges. It will be necessary to closely interact with agro-processing industry for effective value-addition of agro-based-products.

Engineering Education: There is considerable increase in intake, but the quality of engineers being produced is of great concern. The balance between social sciences, art and literature, economics, science and technology should prove to be a potent hybrid mix to create increasingly better skill sets to meet engineering challenges. India needs to enhance its 'Patent Portfolio'. It is imperative that the industry joins hand with academic and R&D community in the spirit of togetherness and sense of fulfilment. Artisan training and skill development programmes are stepping stones to climb up the ladder of increased opportunities and social status.

Habitat: The housing shortage in both the rural and the urban areas is a big challenge. The problem is further compounded on account of unplanned growth of human settlements and ineffective land management in most of the urban centres. The requirement of future housing will require a paradigm shift in engineering approach, in terms of pre-fabricated building components, partially mechanised construction technologies, cost effective and sustained building materials and disaster resistant designs.

Water and Sanitation: Water resources in India suffer from major spatial variations. The magnitude of meeting the targets of social and technological water in India is very large. The remediation of industrial wastewater is a major multidisciplinary engineering challenge. The present state of water supply and sanitation has an adverse impact on the public health. Engineering solutions to these problems are possible interfacing medicine-biology-engineering.

Energy: India has to make a transition to low carbon energy economy at a sustainable speed. India's nuclear energy scenario is encouraging; the country is endowed with outstanding nuclear engineering expertise in pressurized heavy water, fast breeder and more advanced versions of reactors. The Indian power grid system is one of the most extensive in its geographical coverage. The challenges include creation of substantial engineering knowledge base, improving the efficiency of energy end use systems and devices and development of sustainable hybrid energy systems leading to a smooth low carbon energy transition.

Healthcare: The Indian health care sector is changing fast. It includes revamping of existing medical facilities, emergence of wearable devices industry, rising trends in telemedicine and remote diagnosis, and large scale private sector entry in medical education. A proliferation of new start-ups and much greater convergence between medical, information technology and engineering professions is anticipated.

Mobility and Connectivity: The meaning of mobility and connectivity is changing fast due to the changing expectations of the general public. Public understanding of engineering innovations in the areas of mobility and connectivity, and thus the adoption of new technologies, needs timely update.

Indian road infrastructure needs advanced engineering planning and huge public investment to meet the requirements of a multimodal transport system. Smart engineering concepts are needed for fast and efficient maintenance of Indian roads. Indian Railways is one of the largest railway networks in the world. Another development of significance will be the high speed rail facility envisaged along certain long routes. Metro Rail is a potential growth area for many urban centres.

Indian automobile sector has restructured and aligned itself to global markets by absorbing latest technologies. The government has prescribed the automobile vehicle emission standards as per the international emission levels. Lack of well-engineered disposal systems for used automobiles are some areas of concerns.

India has an extensive network of inland waterway sources. The freight and passenger transport through this mode, however, is highly underutilized. Efficient and widespread inland water transportation is vital for India's development. Engineering plays a crucial role in ocean management, ports logistics, ship building, supply chain management, and ship management. Engineering challenges in this field mainly concern maximization of transportation efficiency, new material development, structural design and improved propulsion systems and development of highly efficient, integrated and reliable transportation vehicles.

Indian aviation market is large. In recent years infrastructure facilities are being added to major airports in India. Many private companies have entered the fray. The design, development and manufacture of aircraft call for multidisciplinary aerospace engineering knowledge of high order. With the access to new knowledge on MEMS, material and nanotechnologies, composite materials, intelligent devices and modelling and simulation platforms, India is well placed to design, develop and manufacture passenger aircrafts in the foreseeable future.

India's space programme provides data for agriculture, water resources, urban development, mineral prospecting, ocean resources and disaster management, and many other sectors. Satellite supported tele-education programmes provide distance education, digital interactive classrooms, multimedia and non-formal school and college education courses in India. India is capable of offering globally competitive launch services.

Safety and Security: The natural disasters like floods, tsunami, cyclones, earthquakes, droughts etc., pose major threats to India. The country requires robust and technology backed timely warning systems and effective disaster mitigation measures. Detecting emergency threats due to disease outbreaks through big data mining and monitoring is vital for the Indian society.

Cyberspace security and management is a big area of engineering intervention. Indian Engineering capabilities in information and network security, disaster recovery and end user education have to be upgraded to face the challenges of unpredictable cyber risks. India has come a long way in countering terrorism, but still, it is not free from major threats. One of the big challenges is how to prevent terrorists from employing sophisticated methodologies.

The technology road map the INAE envisions for the country for the immediate future include: Novel materials, particularly for the strategic sectors like Defense, Atomic energy and Space; Energy transition to renewable energy resources; Upgradation of national infrastructure, including smart cities and e-governance; Cyber-physical systems, and that includes robotics and automation,

Internet of Things, cloud computing, 3D printing, machine learning and data analytics, drones, predictive asset maintenance, blockchain technology, digital platforms for integrated designs, etc.

4

Convergence of Engineering and the Science of Life

There is no better engineering than life.

What is life – is a perennial question. Life is defined in different ways depending upon who is defining it. Biologists define life using a list of characteristics, such as metabolism, evolution, homeostasis, constant transformation, genetic information, etc. Life has been defined as self-maintaining and self-producing systems. Physicist Erwin Schrödinger (1), argued that living organisms escape the destiny of disordered equilibrium because they are able to take up 'orderliness' from their environment. Philosophers have defined life from different angles. They say that life can only be understood fully from an 'insider' perspective.

Life is not a random collection of extinct molecules. It is a process of dynamic renewal; everything in life is constantly turning over and renewed. Life exists in diverse forms. Each form is specialised to survive and reproduce in its particular niche.

One understands life from one's own perspective. For some, life is an 'objects in flux' that constantly exchanges matter with the environment. It is systematic material exchange into an organized body. Then there is a view that says that all living organisms deserve moral consideration. All living organisms, however, do not have equal moral significance; vertebrates, for various reasons, deserve more moral consideration than would bacteria or plants. Then there are myriad religious ways to define life. The simplest is "God created the world and all living organisms." No further questions, accept or reject the hypothesis.

In this age of 'biologism', biologists are trying to explain almost everything related to life through biology. In order to do this, biology is very closely interacting with chemistry, physics, mathematics, engineering, psychology, sociology and economics. There is thus a major shift in the way life processes are being studied and understood. Life scientists and technologists are now trying to understand the life processes more quantitatively. The purpose is to understand intricate dynamic behaviour of individual components so as to better appreciate their integration with the whole.

Come post-genome era, not only hopes and aspirations, but requirement for the development of quantitative biology has necessitated effective interfacing between biology and computational sciences. There is growing relevance of system biology in studying the behaviour of biological ensemble. Biological systems are being designed to perform a specific function using synthetic biology concepts. It seeks to construct biological components, such as genetic circuits, metabolic pathways, parts of enzymes as per the required design specifications. Such designed components can be assembled into larger integrated systems to resolve specific problems in hand.

While researchers are using a combination of DNA synthesis, databases and protein science in creating powerful molecular tools, they have been successful in making self-replicating viruses to understand antiviral immunity better. They are using sequences from bacteria, bacteriophages, jellyfish, and the common cold virus to design the self-replicating virus. They are doing this to understand the mechanisms that specifically recognise and disable intracellular pathogens.

As synthetic biology promises to build biological systems from the scratch, it includes redesigning existing genome for a new function. For example, some groups are developing a synthetic version of the *Mycoplasma genitalium* genome that has been stripped down to the absolute minimum number of genes required to support independent life. The goal of this "minimum genome project" is to build a simplified microbial platform to which new genes can be added, creating synthetic organisms with known characteristics and functionality. Protein design, systematically altering the genes that code for certain proteins to achieve desired modifications in protein stability and function, is another application, all-together applied for orchestrating synthetic biology.

At a time when biologists were reading the genetic code, synthetic biologists have made immense leaps and bounds to bring in the early stages of writing the code of life. They are adding new letters, rearranging them into new genetic networks, putting them into an artificial 'chassis' to go forth and multiply.

Synthetic biologists are willing to wait to define life until the time they are able to build living entities. "To assert that life per se cannot be synthesized does not amount to claiming that the products of synthetic biology are not alive, it simply calls into question where 'life' in these entities 'came from'. Most synthetic biologists produce new organisms by introducing genetic elements into existing living organisms. In these cases, we might argue that life was already present in the source organisms and is thus not synthetically produced." (2)

Engineering Biology

'Biological carbon' is becoming the 'Silicon' of this century. In the age of biology many new industries are evolving. In this scenario life scientists need engineers, and at the same time, more and more engineers are responding to the revolutions made in the field of life sciences. Both engineers and biologists have begun to understand that potential is one thing, and to taste the fruits of potential is another. It is where the challenges lie for both engineers and biologists.

Biology and engineering have traditionally represented completely different departments, fields, career paths, even philosophies. But of late, these pursuits have begun to merge in several different ways. There is optimism in the world of life technologies. In order to retain this optimism, engineering and life sciences are coming closer to each other. Life sciences are now so intimately connected with almost all engineering disciplines. One such classical example could be attributed to an emerging discipline called 'Mechanobiology' which coalesces biology and engineering, focussing on the mechanical properties of major components in organisms, such as cells and tissues, and how they contribute to cell differentiation, physiology, and development of diseases. The finite state of diseases and the phenotypes are heralded by understanding the transducing properties of nucleus, the genome architecture and organization.

It is important to understand the behaviour of biological ensembles, and how to design biological systems to perform a given function. Engineers are expected to use advanced engineering tools and methods, and at the same time get familiar with the tools of modern biology. As life is assuming a new meaning, we ask, how best can we address the problems of life sciences to the engineering students? How much biology is enough for engineering students? How do we simplify biology for engineers?

An engineer's approach to understand complex biological systems is different. Accordingly, life sciences could be taught differently to engineers. There is a view which says that "function-based approach with the idea of nature as the designer, and evolution as the design tool" should be followed to teach life sciences to engineers. It means "start with how it works, and then talk about the parts in the whole."

The first revolution in the biological sciences was due to the developments in molecular and cellular biology. This revolution began with the discovery of the structure of DNA. This was followed by developments in genetic engineering. The 'cut and paste technology' allowed researchers to explore inner working of the cells. Thanks to the CRISPR-Cas9 systems, which has revolutionised the way genome is edited. These technologies have enabled researchers to understand diseases at the level of molecular 'hardware' inside the cells.

Genomics is responsible for the second revolution in biology. The purpose of genomic studies is to look into an organism's entire genome. The studies include DNA sequencing, identifying the location of discrete genes, and understanding the inter-genomic phenomena. The second revolution enabled researchers to understand the 'software' driving the cellular processes. Due to this revolution, it was possible to sequence the human genome, as well as the genomes of many other organisms. Due to these developments, many interdisciplinary areas, viz, bioinformatics and computational biology, synthetic and systems biology, nanobiology, biomaterials, and tissue engineering have emerged, thus paving the way for the third revolution.

A revolution of sorts is also taking place in the field of biomedical engineering, where the major concern is to establish an interface between man and machine that are relevant for healthcare and medicine. While a biomedical engineer studies biomechanics at the molecular, cell, organ, and whole-body levels, the cell engineering has extended its boundary beyond the cells. By and large, not only has it included the use of the cells and part of the cells to build structures, such as tissue and ultimately the whole organs, but allowed researchers to use it for plethora of applications in engineering principles, chemistry, nanotechnology, and material science to control cell behaviour. The major application of biomedical engineering is in the areas of biomaterials and tissue engineering. It develops drug delivery platforms, technologies for imaging the body and its components, devices for replacing neurological function, biomaterials for use in the body, organ-replacement systems, artificial blood, etc.

With the growing interface between biology and engineering, the biological function is being seen as 'biomolecular machines', and its regulation as 'biomolecular circuits'. How best to address biological engineering problems? Drew Endy (3) has some suggestions to transform biology in the way integrated-circuit design has transformed computing.

An engineer uses 'nuts and bolts' to assemble a thing. He doesn't do research when he applies these tools. He is confident that the tools he is using are the right ones. This confidence is lacking in biological engineering. 'Biological nuts and bolts', and 'cut and paste technology' need better standardization, as this will make combination of parts made by different manufacturers easy.

One view is that "Genetic engineering techniques are abysmally primitive, akin to swapping random parts between random cars to produce a better car". Endy thinks that "biological engineering community will have to develop formal, widely used standards for most classes of basic biological functions". It means "promulgation of standards that support the definition, description and characterization of the basic biological parts as well as standard conditions that support the use of parts in combination and overall system operation".

Decoupling, "to separate a complicated problem into many simpler problems that can be worked on independently, such that the resulting work can eventually be combined to produce a functioning whole" is another idea that could be helpful. The problems of biological system complexity can be managed by allowing individuals to work at any one level of complexity without regard for the details that define other levels, and also allowing the principled exchange of limited information across levels.

Convergence of life, physical and engineering sciences is being hailed as the third revolution in biological sciences. According to a document prepared by the MIT (4) , this convergence will be the emerging paradigm for how medical, energy, food, climate, and water research will be conducted in the future. The idea of convergence is to bring closer the technical tools as well as the 'disciplined design approach' traditional to engineering and physics, and apply them to life sciences. As the document indicates, convergence is a two-way street. Both the sides seem to be in win-win situation. If engineering and physical sciences are transforming biological sciences, biological models are simultaneously transforming engineering and physical sciences.

Convergence is an 'intellectual cross-pollination' concept. As a result of convergence, several exciting discoveries are shaping into realities. Convergence technologies in biomedical research includes developing nanoscale drug delivery device with an aim to target only the cancerous cells so that harmful effects of chemotherapy can be avoided. This is possible through the convergence of material science, engineering, chemistry and biology. Another interesting area is developing computational models to understand how the body's immune response operates. Yet another area is prevention of blindness through high-tech imaging technology. With the new and fast optical coherence topography (OCT) based imaging technology, it is now possible to spot even the most subtle changes in the retina. The convergence of neuroscience, electrical and computer engineering, and chemistry has made it possible to design brain grafts for treating brain disorders and injury. Various other innovative ideas are likely to come to clinical stages soon. This is possible due to the coming together of various emerging scientific disciplines. In view of this, one of the recommendations of the Convergence document is "Educate, expand, and support the next generation of convergence researchers."

There are many challenges that need to be met both for engineers and biologists, if they want to come closer to each other to understand and appreciate the needs and capabilities of each other. In this techno-bulged scenario, tools of creation are proliferating. Engineers are possessing the ability to innovate and integrate their knowledge across a broad spectrum of disciplines. New technologies and computing power are transforming biology from a "soft" science focused on description to a "hard" science focused on quantifying, predicting, and controlling its properties. It is said that engineers and biologists of the future will be 'T-shaped thinkers', deep in one field and a working fluency in several other fields.

The 'Omics' Landscape

In the post-genome world and in the changing life sciences scenario the need of integration between life sciences and informatics has become essential. This need has paved way for the growth of the 'omics' world. Imagine 'life' cut into small biological pieces. To study the small pieces individually as well as an integrated whole, one would need to study, besides biosciences, informatics, mathematics and statistics.

Omics refers to comprehensive analysis of biological systems. The word 'ome' means many. Genome, the most fundamental part of the 'omics' world, keeps track of the entire genetic makeup of an organism. Genomics, the study of genes, individually and collectively, examines the molecular mechanisms and the interplay of genetic and environmental factors. Due to the advances in structural genomics we have known genome maps of several disease causing bacteria, malarial parasite and plant pathogens. This information is useful to understand disease formation and also for designing and developing drugs. The study of functional genomics tells us how genes are regulated, and how in the presence of one gene other genes work. What make a difference are not the number but the composition, structure, and function of the genes.

The genetic network is regulated by various other 'omics'. Pharmacogenomics studies variations in the human genome in response to medications. It is a way to design drugs, tailor-made for the individual, based on the patient's genetic makeup. Evolutionary genomics studies how genomes change over time. Proteomics is the systematic study of proteins, an important component of structure, function and regulation of biological systems. Gene expression is protein production. Besides generating protein maps, proteomics includes protein expression profiling, post-translational modifications, and protein-protein interactions. Transcriptomics is the study of the transcriptome, the set of messenger RNA (mRNA) molecules or 'transcripts' produced in a species or individual. Metabolomics studies the complete metabolic response of an organism to an environmental stimulus or genetic modification. Systeomics does the integration of various omics-based sciences.

The deliberate modification of the genome is known as designed genomics. It involves targeted, specific modification of the genetic information of living organisms. The technologies that were earlier developed for the insertion of a gene into a living cell were found to have certain limitations. Apart from reproducibility, random nature of the insertion of the new sequence into the genome and blind positioning of the new gene that may disturb the functioning of other genes were a few limitations. They could also trigger the process of 'cancerisation'.

The major advantage of designed genomics is precision and reproducibility. Due to the recent developments in the area of genome engineering, it is now possible to enhance the de novo assemblies of DNA at a reduced cost. "These developments promise genetic engineering with unprecedented levels of design originality and offer new avenues to expand both our understanding of the biological world, and the diversity of applications for societal benefit", write P A Carr and G M Church (5).

Designed genomics, a very active field of research at present, involves introducing a gene into a chromosome to obtain a new function. The insertion of gene is also to compensate for a defective gene, particularly by making it possible to manufacture a functional protein if the protein produced by the patient is defective. Designed genomics also involves inactivation or 'knock-out'. The purpose of inactivation or 'knock-out' is to know the function of a gene by observing the anomalies that occur as a result of its inactivation. Correction aims to remove and replace a defective gene sequence with a functional sequence. Another aspect of designed genomics is 'correction'. It is intended to correct short sequences (sometimes just a few nucleotides), such as in the case of sickle cell anaemia.

Designed genomics, a technology based on gene targeting, has a wide range of possible applications, like the correction of a gene carrying a harmful mutation, the production of therapeutic proteins, and the development of new generations of genetically modified plants. Gene targeting technology can bring precise changes into the protein coding part of a gene if one knows the mutation that causes the disease. What do designer genomes have that the traditional genomes don't? Given the fact that there is genetic variation centric to the genomes, the mutations harboured can be 'repaired' and set for a change. As these genomes burgeon, one question still eludes the scientists, would the designer genomes have 'acclimatibility' when compared to their peers or traditional genomes?

The use of gene targeting to evaluate the function of genes in the living mouse is now a routine procedure. A mouse could also be used to produce a human version of protein. We now know that more than 90 per cent of the genes have a function shared between a mouse and a man. More than 10,000 genes have been targeted in mice. More than 500 different models of human diseases, including models for hypertension, atherosclerosis, cancer, diabetes and cystic fibrosis have been produced by gene targeting. Gene targeting has also been used to understand the role of individual genes in embryonic development, adult physiology and aging.

Mario Capecchi, Martin Evans and Oliver Smithies (6) have discovered the principles for introducing specific gene modifications in mice by the use of embryonic stem cells. Capecchi demonstrated that defective genes could be repaired by homologous recombination with the incoming DNA. Homologous recombination is a type of genetic recombination in which nucleotide sequences are exchanged between two similar or identical molecules of DNA. Capecchi discovered genes crucial for mammalian organ development and the body plan in general. Capecchi's research has shed light on the causes of several birth defects. Evans discovered that chromosomally normal cell cultures could be established directly from early mouse embryos, referred to as embryonic stem (ES) cells. He developed mouse models for human diseases. Smithies discovered that all genes may be accessible to modification by homologous recombination. Smithies used gene targeting to develop mouse models for inherited diseases such as cystic fibrosis and the blood disease thalassemia. He also developed numerous mouse models for common human diseases such as hypertension and atherosclerosis.

Designed genomic modification has several advantages for introducing mutations into mice, writes Capechchi. The investigator can choose which genetic locus to mutate. The technique takes full advantage of all the resources provided by the known sequences of the mouse and human genomes. The investigator has complete control of how to modulate the chosen genetic locus. Gene targeting in mice has pervaded all fields of biomedicine, and as Capecchi concludes "The repertoire of biological phenomena that can be studied through the use of gene targeting is only limited by the imagination of the investigator."

One misconception about omics-based sciences is that it will replace lab work. That is not true. The omics-based sciences will not replace lab work. It, however, has the potential to reduce number of experiments, thereby reducing time and money. It will help researchers to plan their experiments more efficiently by eliminating false positives, which consumes most of the time and money of the researcher.

The commercial 'omics' landscape seems rosy. Almost all IT majors are involved in this activity (data warehousing, integration and application support, mainly at the pharma companies). There are also 'pure play' omics companies with domain expertise in discovery informatics outsourcing and solution architecture, data and system integration and control provision, and tools provider.

The major challenge for the industry is to anticipate what will be the next useful data sources and analysis tools. The key to selling products in this sector is in making sure that your company's solution clearly fits an established need and integrates seamlessly with other products already in place. The big question the omics industry face is whether to purchase content or technologies from commercial vendors or to develop tools in-house to meet their specific needs. One estimate shows that biotech companies are outsourcing 60% of their bioinformatics work. The 'omics' industry expects a big jump due to projected scenario of increased R&D expenditure in the pharmaceutical companies.

Synthetic and Systems Biology

"For the first time, God has competition", writes Nature. God, perhaps, is happy creating his own competitor. He has created synthetic biologists who are making attempts to redesign a world, possibly a better world. Molecular biologists, chemists, computer scientists and engineers have joined hands to create a synthetic world.

To build this world they are evolving tools and techniques. They want to build bio-bricks to construct new life forms. Synthetic biologists are mastering the art of gene manipulation and trying to reconfigure the metabolic pathways of cells to perform entirely new functions. For obvious reasons they want to create the simplest form of life, as the simplest form of life has minimum things that are essential for a life to exist.

'The whole is greater than the sum of its parts', sums up the basic tenet of systems biology. It is a holistic approach integrating many scientific disciplines to predict how these systems change over time and under varying conditions. It is a way to create the potential for entirely new kinds of explorations, and new kinds of explorations require novel analytical tools. Systems biology is, in a sense, "classical physiology taken to a new level of complexity and detail." Systems biology is often referred to as the "Network of Networks."

Our body communicates on multiple scales. At the biological level, it is genes to molecular networks to cellular networks to organ networks. At the social level, an individual networks with other individuals. Systems biology looks at these networks across scales to integrate behaviours at different levels. Our body is a complex machine. Organization of its hierarchical structure is a multiscale phenomenon, and is governed at multiple levels, spatio-temporal besides being genetic, and influenced by the environment. Although quite a number of sophisticated computational models and simulations represent integral parts of systems biology. To study systems biology, a systematic measurement such as genomics, bioinformatics and proteomics, and mathematical and computational models are used to describe and predict dynamical behaviour.

Systems biology does not accord reductionism in taking the pieces apart, but requires understanding of engineering principles, nonlinear systems analysis, network theory, abstract mathematics, nonlinear thermodynamics, physics, chemistry, and biology. Howsoever imperfect, different analogies have been put forward for describing systems biology, like, a chemical plant, an electronic circuit diagram, and so forth and so on.

On the other hand, systems biology suffers from various kinds of paradoxes. It has a 'Scale' paradox: genome vis-à-vis small scale networks. It also suffers from 'Discipline' paradox: biological vs physical. It is bothered by the paradoxes of 'Method' (computational vs experimental) and 'Analysis'(deterministic vs probabilistic). Though systems biology differs from synthetic biology (systems biology studies natural biological systems as a whole, whereas synthetic biology aims to build novel and artificial biological parts, devices and systems), both use many of the similar methods and approaches. Systems biology uses these approaches to better understand the inner-workings of life, whereas synthetic biology's emphasis is on engineering living systems to behave in specified ways. Synthetic and systems biology are closely dependant on each other; "the complexity imbued by interactions among components is vitally important to the proper functioning of the system-to-be."

What will happen if the tools and techniques of synthetic biology become accessible to common people? As someone observed, "Designing genomes will be a personal thing, a new art form as creative as painting or sculpture. Few of the new creations will be masterpieces, but a great many will bring joy to their creators and variety to our fauna and flora." There is also a possibility of synthetic creations creating havoc in the natural ecosystem. "We might be able to create a bio-unit that looks and performs perfectly in the lab, but we cannot control how ecology and evolution might rewire our synthetic unit in an ecosystem, nor can we predict how that synthetic unit might rewire the ecosystem and its inhabitants," writes Seirian Sumner (7). The risks are many and there is danger of the natural world becoming unnatural.

Genome Editing

Genome editing is a powerful new tool for making precise additions, deletions, and alterations to the genome. Editing the gene for treatment or prevention of disease and disability has great potential. The possibilities include (1) restoring normal function in diseased organs by editing somatic cells, (2) to preventing genetic diseases in future children and their descendants by editing the human germline, (3) human enhancement.

Genome-editing research using somatic cells can potentially facilitate the ability to develop better interventions for affected people. Genome editing of germline cells can help in supporting advances in such areas as regenerative medicine and fertility treatment. Use of genome-editing for "human enhancement" is a controversial subject that needs public and governmental intervention.

The idea of making genetic changes to somatic cells is not new. A number of ways have been suggested for somatic genome-editing therapies. These include removing relevant cells (such as blood or bone marrow cells) from a person's body, making specific genetic changes, and then returning the cells to that same individual. It could also be performed directly in the body by injecting a genome-editing tool into the bloodstream or target organ. *In vivo* gene-editing strategies have a few technical challenges (like tool not finding the target resulting in little or no health benefit) that are yet to be resolved, but a few clinical trials of *in vivo* editing strategies are already under way. The effects of genome editing of somatic cells are limited to treated individuals and are not inherited by their offspring.

Genome-editing of an individual's germline cells in human systems is a major technical challenge. With tremendous interest in this area of research, thousands of inherited diseases caused by single-gene-mutation are explored. Germline genome "could provide some families with their best or most acceptable option for averting disease transmission, either because existing technologies, such as prenatal or preimplantation genetic diagnosis, will not work in some cases or because the existing technologies involve discarding affected embryos or using selective abortion following prenatal diagnosis."

Among a few genome-editing technologies available, CRISPR (clustered regularly interspaced short palindromic repeats) provided the foundation for the development of a system that combines short RNA sequences paired with Cas9 (CRISPR associated protein 9, an RNA-directed nuclease), or with similar nucleases. The CRISPR/Cas9 genome-editing system offers several advantages over previous strategies for making changes to the genome. The system can be readily programmed to edit specific segments of DNA, and can be engineered more easily and cheaply than other methods to generate intended edits in the genome.

Genome editing is a relatively new phenomenon. As is obvious with anything new, it is generating considerable enthusiasm as well as nightmare. As with other medical advances, questions are being raised with respect to genome editing. Questions include: (1) balancing potential benefits against the risk of unintended harms, (2) governing the use of these technologies, (3) how to incorporate societal values into salient clinical and policy considerations, and how to respect the inevitable differences. Public interest and engagement, therefore, is an important part for the success of this endeavour.

Use of genome editing for "enhancing traits and capacities beyond levels considered typical of adequate health" could involve both somatic and germline cells, and thus the uses of the technologies raise questions of fairness, social norms, personal autonomy, and the role of the government. It raises the controversial question related to "enhancement" and "normal". The pertinent question that is being asked is - should we not restrict the application of genome editing for the purposes of treatment or prevention of disease and disability only for the time being.

Bioprocessing and Biomanufacturing

Biomanufacturing faces various engineering challenges. It undergoes different stages of development. The first is discovery and 'proof-of-concept' stage. Here the bench scale data is transferred to the process and analytical group. This group examines the efficacy of the process/product for scale-up, product quality and regulatory matters. Changes in the process or materials, if required, are examined in terms of safety and efficacy of the product(s). If the results are positive, both technically and business-wise, decision is taken to go further and apply for the approval and licensing. Proof-of-concept *in vitro* and *in vivo* must be validated before any drug can move from discovery to a clinical development program.

Contract Manufacturing Organizations (CMOs) help develop and manufacture the molecule as well as offer quality and regulatory support. In the preclinical phase, the molecule's phenotypic, genotypic, and biochemical profiles are determined. These include the shape of the molecule, amino acid sequence, isoelectric point (pI), molecule's mechanism of action, its potential bioactivity or availability, and any possible toxicity issues.

One big problem with biotech industry is value creation. "The truth is that some wonderful science has been developed by the contemporary crop of technology platform companies; however, they all essentially address nothing more than a thin slice of the drug-discovery continuum," writes Stelios Papadopoulos (8). Papadopoulos, asks, "What happens then to the hundred or so early-stage technology platform companies, both private and public?"

According to some pioneers, the key drivers of success are truly outstanding leadership, scientific talent, and aggressive capital raising. The challenge is "how to manage the increasing number of relationships and how to optimize technologies sourced from different companies that may, on occasion, be largely incompatible."

Guido Lanza (9) writes (for the pharma industry driven by process innovation) that the increasing R&D costs and a declining number of new chemical entities' approvals (due to inefficiency of discovery and the attrition of new chemical entities through development failures) is creating an atmosphere for the platform companies to resurface. Guido Lanza has listed a few criteria that all successful platform companies must meet:

(1) Technologies applicable only to a small set of proteins, or docking technologies that only generate hints or explanations for medicinal chemists to use, are unlikely to form the basis of valuable platforms;.

(2) If the technology is an add-on to the existing process, then it may form the basis of a service business, not a successful platform company.

(3) Those technologies requiring substantial initial capital to drive programs forward may not be suitable for a successful platform technology.

(4) Make the choice between building a horizontal business by partnering with companies and building a vertical business by picking one or more product opportunities to develop.

Biomanufacturing facilities for human therapeutics require large capital investment and high operating costs, and demands highly skilled workforce to run the chain that includes discovery, validated manufacturing processes and drug approval. "Keeping this in view, modular systems are being used for multiple products in a single site. For most biologics 2000 L is a good size."

Irene Berner, Susan Dexter and Parrish Galliher (10) list the following categories of enterprises that are either considering or already moving to build biomanufacturing capacity in new geographies: Large, incumbent biopharmaceutical and vaccine companies, Established midsized biotechnology companies, New contract manufacturers, Government-backed initiatives focused on biodefence and local populations, Emerging biotechnology companies working on new drugs and biosimilar development, Small-molecule companies seeking better margins by transitioning to biologics. Some are planning to scale down their mega-production facilities in favour of smaller, more flexible plants that are built closer to global customers. Some small-molecule companies that face dwindling pipelines and shrinking margins for existing products are making large investments in biologics research, development, and production.

The biotech industry brings with it a new resource base, a new set of transforming technologies, new forms of commercial protection, a global trading market, a new supporting sociology, a new communication tool to organize economic activity at the genetic level. The biotech R&D is science driven and competitive, and it leads to lengthy market lead times and patent related issues, and, therefore, R&D costs are high.

Biotech revolution has gone through various phases. In the first phase of industrial biotechnology revolution, emphasis was on research methodologies which changed how we study and view biology. The second phase concentrated on new discoveries, based on new methodologies. In the third phase we saw the flow of innovative products to the market. In the fourth phase, emphasis was on the incorporation of products into daily life.

Biotechnology is now the integral part of society. Emphasis on a particular technology also came in phases. First it was recombinant DNA, then monoclonal antibodies, and now it is the turn of gene splicing leading to genomics and gene therapy. Most products are based on the first and second platform technologies. Products based on third platform technology still have to show their mettle.

The biotech industry thinks it is the 'ultimate technology frontier'. Some doubt the efficacy of biotech revolution. Overestimation in the speed and extent of the impact has created doubts.

As our skills and capabilities develop, the 'world' we see literally shifts. This is exactly what has happened with the biotech industry. Biotech industry wants to rival the growth curve of the industrial age by producing materials through 'living' means, and that too at a pace far exceeding nature's own time frame, and then converting the material so produced into an economic opportunity.

This is one industry which is labelled as 'risky'. Since this technology concerns life, risks accompany hurdles. Risks and hurdles include complex regulatory approval processes and costs, high research costs, and difficulty in the availability of start-up capital. The industry, as they say, is overwhelmed with products in early stages and under-whelmed by products that can be marketed. And, as is well-known, investors are more interested in late-stage discoveries, rather than embryonic technologies.

The industry faces fierce competition. There is the threat of substitute products, existing competitions, and the threat of new entrants. Often one comes across a paradoxical situation — the researchers tend to forget that the purpose of industry is to make money, and the investors believe "we finance businesses, not science projects".

Researchers as well as investors have understood that there is a need to shift technology paradigm. They have understood that R&D investments are often more than capital investment. They are thinking of shifting from one-company-one-technology organization to an enterprise housing multiple technologies. They are thinking of the transition from in-house R&D to cross-industry long-term R&D cooperation. The emphasis is shifting from supply-side technology innovation and development to demand-side R&D effort. More and more, it is being realised that 'blue-sky scientific enquiry' emanating from fundamental human curiosity requires to be better balanced by the needs of wealth creation, employment generation, social justice and care of the environment.

India is blessed with rich biodiversity. It has qualified human resources and good network of research laboratories. It has a developed industrial base. These are a few good things for the growth of Indian biotech sector. The R&D expenditure on biotech needs improvement. The industry needs better capital inflow. The country certainly needs greater cohesion in the language of market place and the language of scientific enquiry for the growth of biotech industry. One of the key success factors for biotech ventures in the country is long-term perspective.

The biotechnology industry in India has immense potential to emerge as a global player. India constitutes around 8 per cent of the total global generics market, by volume. Hybrid seeds, including GM seeds, represent new business opportunities in India based on yield improvement. There is an opportunity in focused R&D and knowledge-based innovation in the field of industrial enzymes. Another area of opportunity is the area of bio-markers and companion diagnostics, which will enable to optimise the benefits of biotech drugs.

Though it is a good idea to generate ideas, it is hard to generate money from ideas. Idea people are available in both the industry and the academia. Idea people from academic institutes are generally inexperienced in various facets of setting up a business, and often lack in handling bureaucratic aspects of building a business. Bioentrepreneurs thus tend to gravitate to biotech clusters in the hope of getting the support of experienced business managers, so as to tap an environment rich in experience—both scientific and executive.

Svea Grieb, Kai Touw and Dan Kopec of Sartorius Biotech take a brief look into the Future of Bioprocessing. The next generation biomanufacturing includes Process Analytical Technology (PAT), Data Analytics and Bioprocess Automation. They envisage several advantages: consistent, high product quality; reduced risk of lost batches and increased process safety; fast and predictive up- and down-scaling; freeing up operators; and account for cell variation from different sources.

Key players of next generation biomanufacturing are flexible, automated skids capable of handling different type of unit operations; abundance of online spectroscopic techniques in both upstream and downstream bioprocessing for the measurement of analytes, cell and product quality; and multivariate data analytics to assess the state of the process, and simulation and modelling of process design and control. Then there are regulatory, logistics and safety issues that have to be solved before automation can really be adopted widely in the bioprocessing industry.

The far-future vision (ten years from now) of Sartorius Biotech is "highly influenced by the industry 4.0 approach and related concepts such as machine learning and the internet of things. .... Processes can be monitored and controlled remotely. Every process will have a digital twin that can be used for process simulation and prediction. More and different data will be gathered and will reside in the cloud, where data analytics can be applied easily to improve processes, regardless of manufacturing location."

A report by The Medicine Maker, sponsored by Merck, on the exploration of the future of bioprocessing observe the following:

(1) "in the not too distant future, continuous processes that drive down cost, PAT technologies that provide real-time process control, and contemporaneous batch disposition, plug and play equipment that allows us to manufacture new products efficiently - all of this will work in concert to make biopharma supply chains tremendously more efficient and to alleviate the impact of demand uncertainty."

(2) "the ideal bioprocess of today would be an inherently batch process that leverages high productivity technologies that ultimately support the evolution to a fully continuous process."

(3) "In the future, I envision a more commoditized manufacturing model as the industry matures. Supply chains as a whole will also experience significant evolution with considerable gains in the reduction of lead times and progression towards an ondemand supply infrastructure to best meet market needs."

(4) "My vision for the future of bioprocessing is having flexible, easily-connected intensification blocks that enable a process to easily meet a particular customer's needs. Every customer is going to have different templates and different objectives, so it is incumbent on us as suppliers to have these building blocks that can seamlessly be integrated, as well as independently of one another, in case a customer is having a problem with one particular part of a process."

(5) "The recent advent of sterile single-use technology provides rapid, low cost implementation."

(6) Pipelines are diversifying and molecules are becoming increasingly complex.

(7) There is significant competition in the marketplace – with similar therapeutic modalities aiming at identical targets with similar modes of action.

(8) Manufacturing platforms need to be agile, able to adapt to specific requirements, extend capacity rapidly, and be responsive – able to react to market demand quickly.

(9) Today's facilities take an average of 5 years to build and commission, but future facilities will target 18 months.

Biology and Technology

The use of life has a long history

Industrial biotechnology is undergoing major changes. The innovation lead technologies include low energy consuming and low waste generating bio-based processes using renewable raw material. This move is necessary to meet the demands of energy, resources, and food of the future world when the state of oil, gas, and coal reserves will not be that satisfactory. The possible solution is to switch from petroleum-based to biomass-based feedstock and use environment friendly and sustainable technologies. Such a shift promise generation of lower and less toxic wastes. Additional advantage of using biomass as feedstock is that it will lower generation of greenhouse gases. The technology jump is possible only if we have better processes, both economically and environmentally. Better processes are possible if we have better organisms.

Bioscientists are discovering novel organisms and biocatalysts from the natural environment in the hope that these organisms can grow and function optimally at relatively extreme levels of acidity, salinity, temperature, or pressure. It is hoped that such 'extremophiles' will mimic the extreme manufacturing conditions of the specific industry. The jump is possible if, in place of conventional catalysts, more effective biocatalysts (enzymes) are used. This will need development of biocatalysts with capabilities to catalyse a broad range of reactions having greater versatility. This will need increase in biocatalyst's thermostability, activity, and solvent compatibility. This will need development of genetically modified organisms.

A big market awaits biopolymers ('green plastics'). Sugar-based biopolymer polylactic acid (PLA) has been welcomed by clothing and packaging industry. Researchers are developing techniques to make natural protein polymers such as spider silk by microbial fermentation. Speciality and fine chemicals, because of their higher value, are the prime targets of industrial biotechnology. The bio-based process is reported to reduce hazardous waste generation by over 70 per cent, water to waste discharge by over 66 per cent, air emission by 50 per cent, and production cost by 50 per cent.

It is difficult for a bioprocess to compete directly with large-volume chemicals produced from common petrochemical feedstocks in fully depreciated assets. Moreover, it is recognized that bioprocesses should be viewed as complementary to thermochemical processes, rather than competing with them. In the future, many chemicals will be produced by a combination of biological and conventional chemical synthetic steps.

Due to changing requirements in the chemical industry, human resources trend is also changing. The chemical industries are now hiring increasing number of life scientists, and chemists are embracing biocatalysts as tools for new synthesis. The innovative biotechnologies are using cheaper raw materials, but the conversion cost of raw material to final product is still not competitive. The great challenge for industrial biotechnology is to develop clean technology at a much reduced cost.

The synthesis of a designer molecule, says J D Keasling (11), is like building refineries and other chemical factories from unit operations.The synthesis of designer molecules is possible using tailor-made microorganisms and engineered metabolic pathways. The synthesis requires the design of pathways, enzymes and their genetic control. The transformation of substrate to product of choice is possible by transferring product-specific enzymes or entire metabolic pathways from a rare or genetically intractable organism to a host organism.

The synthesis of desired product is also possible by combining enzymes or pathways from different hosts into a single microorganism. Another approach of making designer molecules is to use engineered enzymes with new functions. In the event, an enzyme does not exist for a particular reaction or set of reactions, one would have to use computer-aided design software to design the desired enzyme, and then introduce it into the host system.

Computers are used for designing designer cells and their genetic control systems. They are designed to control expression of all genes at the correct time, and at appropriate levels. The designer cells are also designed to deal with genetic control system failure, if any transient failure occurs during the production process. Based on the requirements of the cell, a chromosome could be either designed and constructed, or obtained from a commercial vendor.

"One can even envision a day when cell manufacturing is done by different companies, each specialising in certain aspects of the synthesis — one company constructs the chromosome, one company builds the membrane and cell wall (the bag), one company fills the bag with the basic molecules needed to boot up the cell," is how Keasling visualises the future of designer catalysts and designer molecules.

The impressive list of biotechnology-based future products includes plants, animals, microbes, and synthetic organisms that are being engineered for deliberate release in an open environment, contained products and platforms (12). Bioengineered products for open environment includes agricultural crops, cleaning up contaminated sites with engineered microbes, replacing animal-derived meat with meat cultured from animal cells, and controlling invasive species through gene drives. The US National Academies document says that "future biotechnology products that are produced in contained environments are more likely to be microbial based or synthetically based rather than based on an animal or plant host." For obvious reasons advanced molecular tool boxes will be in great demand. Biotechnology platforms include 'dry' and 'wet' labs; wet labs such as DNA/RNA, enzymes, vectors, cloning kits, cells, library prep kits, and sequencing prep kits; dry labs such as vector drawing software, computer-aided design software, primer calculation software, and informatics tools. The document recognizes that social trends impact technology trends and therefore there are issues of competing interests, risks, and benefits regarding future biotechnology products. From time to time society has raised concerns over safety and ethics.

Life technologies is filled with so many ideas. Many of them have demonstrated their worth at the laboratory scale. A few have been validated at the pilot scale. The road to commercialization for the most is not easy. It is a 'hard-to-please' industry. It needs to fill many gaps. It needs to construct many bridges. It needs to cross many 'valleys of death'. It needs to know how the new technology fits in to an old society. One of the problems with idea people is that these people are overconfident about the usefulness of their ideas. These people feel convinced "that they are de facto going to succeed". But sadly, this belief often doesn't bear fruits.

It needs to bridge the knowledge gaps. It needs to bridge the gap that arises due to overflow of ideas. It needs to bridge research/technology/manufacturing gap. It needs to bridge the gap between scientists/farmers/ market leaders. It needs to bridge the gap between nano, micro, and macro technology. It needs to bridge the skill gap. It needs to build the gap between hope and aspiration, expectation and reality. It needs to build the gap between theory and practice. It needs to bridge the gap between class room (campus) and the real world (corporation). It needs to bridge the funding gaps. It needs to bridge the gap between Public/Private. It needs to bridge the gap between research ability and entrepreneurial ability. It needs to bridge the gap between employment and employability. It needs to bridge the gap between the present and the future. It needs to bridge the gap between individual and corporate responsibility. It needs to bridge the gap between company strategy and strategic alignment.

We can't contain the scope and speed of biotechnology. We can't even keep track of our own narrow field of interest. We are supposed to know what others are doing, more than what we ourselves are doing. We follow the 'trend'. We need to bridge the gap between competition and cooperation. We need to bridge the gap between intelligence and information.

As Nature writes in one of its issues devoted to biotech industry (13), "In the biotechnology industry, leverage is everything: people, financial resources, technology, and patents are all leveraged for success. A key contributor to successful leverage efforts is a very simple one: typically, whoever knows more in advance and responds to that knowledge quickly wins. This is where competitive business intelligence systems come in."

The key drivers for the biotechnology products include genome-engineering technologies; standardization of biological parts; and increasingly rapid design-build-test learn cycles. The developments in DNA sequencing, synthesis, and editing has enabled targeted modification of DNA sequences—such as insertions, deletions, and site-specific replacements of DNA bases—in a variety of organisms. Understanding of how RNA interference (RNAi) silences gene expression are creating opportunities to create new products. These are helpful in synthesizing a wide variety of organisms and new genetic constructs to modify the function of living organisms. One of the major challenges of biotech-based products is the reproducibility issues at the scale up stage. Often engineered microbes stop producing the desired product or the production rates fluctuate. Another need is proper documentation of biological components, including description and performance characteristics. Another major challenge is lengthy design-build-test learn cycle.

Biotechnology is a way to solve various problems facing humanity. It affects different societies differently. Some societies perceive it as a threat; threats to human health, the environment, biodiversity, and resource accessibility. It is important to appraise the members of such societies the benefits in comparison to the concerns.

Some other major achievements in biotech include gene therapy, the cell atlas, diagnosing and treating disease using the microbiome, absorbable heart stent, smart contact lense, cancer spit test, etc.

Gene therapy is an attempt to treat the patient by replacing the gene responsible for destroying the immune system (14). A boy suffering from immunological disorder gets an infusion of the therapy in his veins, and becomes a normal boy thereafter. It is an elegant way to treat those suffering from an error in a single gene. After the initial disappointment, the method is now a success. It can cure devastating genetic disorders. There is hope; 40-50 gene therapy clinical trials are underway. Researchers are hopeful of treating diseases, like Alzheimer's, diabetes, heart failure, and cancer. Of course, there are still many challenges to overcome.

The Cell Atlas is insights into high-resolution spatio-temporal distribution of proteins within human cells (15). The Cell Atlas currently covers the genes for which there are available antibodies. "It offers a database for exploring details of individual genes and proteins of interest, as well as systematically analysing transcriptomes and proteomes in broader contexts, in order to increase our understanding of human cells. It will provide scientists a sophisticated new model of biology that could speed the search for drugs."

Absorbable heart stent is an alternative to the small metal mesh or scaffold (called stents) inserted into the blocked artery to widen the vessel and restore blood flow to treat coronary artery disease and angina. The regular stents do not remain good for longer periods. Sometimes, scar tissue forms within the stent, causing the artery to narrow again and necessitating a repeat procedure. Absorbable heart stent is relatively safer compared to metal stents. "The device releases a drug called everolimus that limits the growth of scar tissue; the underlying scaffold is designed to fully dissolve into the body after about three years." It is made of a plastic-like polymer and works much like surgical sutures that dissolve. "For a younger patient, typically age 65 or younger, with big arteries, and a lot of heart muscle, these devices can be really optimal," says Gregg W. Stone (16).

Soft biosensing contact lenses are being developed to detect glucose levels in patients with diabetes (17). This smart contact lens, it is hoped, can monitor glucose levels from tears in the eye, and it is hoped it can lead to resolve a host of diabetes complications. The research team uses electrodes made up of highly stretchable and transparent materials to solve contact lens discomfort issues. The prototype has been used on rabbits that showed no visible discomforts. This non-invasive healthcare unit offers substantial promise, say the researchers.

A new technique to detect cancer from a drop of person's saliva is being developed by a group of researchers from the University of California at Los Angeles (18). Known as 'liquid biopsy' the saliva test, if successfully deployed, could save valuable time and enable doctors to detect and treat cancer earlier than currently possible. The test is based on the fact that traces of RNA from cancer cells can be found in saliva. RNA in the sample can tell what sorts of processes are going on inside a cell, including those associated with cancer. The test has been successfully tried on lung cancer patients. One of the big advantages of the test will be detecting multiple type of cancers in a single test.

Life Lessons From Life Sciences

The science of life at different levels and hierarchies has given us helpful insights, not only to organize life, but also to organize an enterprise. It works on the premise that organisations are likely to face situations that living organisms have faced and mastered during their several billion years of evolution.

Isn't it amazing that we start working even before we are born? "To become an embryo, you had to build yourself from a single cell. You had to respire before you had lungs, digest before you had a gut, build bones when you were pulpy, and form orderly arrays of neurons before you knew how to think." We are born before we are fully developed, and that's the reason we need years of intensive care before we can fend for ourselves. Tadpoles, on the other hand, are ready to swim, find food and evade predators the moment they are born. But then, we are humans, not tadpoles. There is a message for organisations in this: Don't treat an organisation like a tadpole, treat it as a human being; give the organisation enough time to find food, fight predators, and learn swimming.

Another important question is the degree of complexity an organism or an organisation can withstand. Complexity is observer dependent; what is simple for one observer can be complex for another? It is thus important to recognise the 'relativity of complexity'. A reasonable guiding factor could be that the system must be stable and persistent, i.e., it must be able to survive. Survival should, thus, dictate the level of complexity an organism or an organisation can withstand.

Life exists at different levels, and the levels are organized in a hierarchical fashion: Molecule – Macromolecule – Cell – Tissue – Organ – Organ System – Organism – Population – Community – Ecosystem. A living being is made up of many atoms and molecules. It lives when these atoms and molecules follow a particular configuration and there is a certain relationship among them.

Life becomes a lump when the desired link between structure and pattern is broken. There are many ways to break these links. Besides the natural way, living beings die as a result of mishaps, such as starvation and injury.

There is another kind of death in which cells die by committing suicide. They do it to ensure proper development of the remaining cells. If they remain in the system, the integrity of the organism gets spoiled. Thus, it is essential to remove them. The pattern of death is so orderly that the process is called 'programmed cell death', also known as apoptosis.

"Life cannot exist without death...apoptosis (programmed cell death) is important for an organism to be able to eliminate unnecessary or damaged cells from its body as it has to generate healthy new cells. Moreover, aberrations in apoptosis are now believed to contribute to many common disorders..."

Programmed cell death (PCD) is a natural process and is essential for our survival. PCD encourages self-destruction of the damaged, diseased, or unwanted cells. Our hand has five fingers, and that is only possible because the cells that lived between them died when we were embryos. PCD ensures a constant turnover of cells in the gut lining and generates our skin's protective outer layer of dead cells. PCD also allows the body to eradicate destructive cells. If there was no PCD, we would face 'runaway cell replication', and that might lead to cancer.

A somewhat similar thing happens in organisations. New kills the old ones. Economist Joseph Schumpeter called this 'creative destruction'. He argued that innovation replaces (destroys) the established enterprises and makes way for new enterprises. In this age of innovation enterprises are running at the speed of Moore's law - "high cost to create, minimal cost to produce, and a winner-take-all environment." This scenario suggests that enterprises must take resources away from the losers, and reallocate them to the winners.

Constructive destruction provides some answers to "why 'built-to-last' enterprises tend to be under-performing and 'upstarts' are over-performing?"

In a discontinuous market, Peter Drucker called it 'age of discontinuity', new entrants are showing promising results. The consumers of wealth, the weak performers, if replaced, can yield better results. Innovations destroy obsolete technologies, only to be assaulted in turn by newer and more efficient rivals.

An organisation can run, even when a part of it is closed. This is also possible in the case of living beings, but they being more integrated and complex, the chances of survival of remaining organisms are comparatively less likely. Programmed cell death teaches us several lessons. "Block apoptosis and development goes awry. Were it not for death, we would not even be born".

There is so much to learn from life sciences for the management of collaborations.

What is better – a smarter group or a diverse group? Studies indicate that diverse group of problem solvers is better than the group of the best and the brightest. "It is like a fruit basket and a shiny apple; the better the individual apple, the better the fruit basket, and the better the other apple, the better the individual apple." In a diverse group each person brings a unique combination of needs, experiences and attributes. Limited diversity is good to boost the self-efficacy of the collaboration. Too much diversity, on the other hand, inflicts conflicts to disturb the collaboration. Moreover, the cost to maintain collaboration depends upon the extent of diversity. Too much diversity costs too much.

Collaboration is like a microbial community, where multiple species are present. These species interact differently; some interactions result in no change in one species due to the presence of the other (neutralism). In some interactions, the presence of one species increases the growth of the other, and vice versa (mutualism). It is possible that the first species is not affected by the presence of the other, but the second species is affected; the second species may enjoy the benefit (commensalism) or may suffer (ammensalism). In competition, one species exert a negative influence on the other. In coalition, we want only positive interactions, but that is too much expectation. Neutralism is rare in natural ecosystem, as well as in collaborations. Mutualism is more commonly occurring phenomenon in nature, as well as in collaboration. In special circumstances (like when collaboration stability is in danger) commensalists, ammensalists, and competitors play their roles to stabilise/destabilise collaboration.

One of the problems in collaboration is introduction of 'special species'. These species are unreasonably flexible, highly demanding, and generally 'uncultivable'. It is thus advisable to be more vigilant while accepting such special species into the fold of collaboration. Then there are 'stressor species'. As the name implies, they get stressed at the smallest pretext. It is important to keep them off-stress, but that is not always possible. Moreover, disease causing microbes (even if small in number) can't be ignored.

Shouldn't we evolve collaboration's major armoury, the common minimum programme, from the concepts of 'minimal genome'. The minimal genome contains only 'essential' genes required for the replication and survival of the novel organism, in a particular environment. One must, therefore, know/determine which genes are essential for basic metabolism and replication, and also know how to provide or create the necessary non-genetic components for successful gene expression. It is good that the minimal genome does not include redundant genes. It is, however, not easy to create a minimal genome; even slight genetic alterations can have far-reaching, unintended consequences.

What collaboration wants is a balance of variability and commonality. We have known from microbial communities that an ideal system comprises of well-characterised individual species that together accomplish functions that are beyond the capabilities of the individual species. The key logic of collaboration should therefore be the identification of such individual species.

A Humane Organization

Ernst Schumacher classified minerals, plants, animals, and humans on the basis of their 'levels of being'. M (mineral) is a non-living entity. Plant is (M+X); 'X' is 'ontological discontinuity' that transforms a non-living entity to a living one. Animal is (M+X+Y); 'Y' is 'consciousness' (animals are conscious in the sense that "they can be knocked unconscious"). "Man has power of life like plants, power of consciousness like animals, and evidently something more." 'Z' is that 'something more'. Schumacher called it 'self-awareness', meaning "man not only is able to think but is able to be aware of his thinking."

Abraham Maslow's human need has five levels. The first level is biological and physiological needs, such as food, shelter and warmth. The second level is need for safety, security, stability, and law and order. The third level is about love, family, affection, relationship and work. The fourth level is about achievement, status, responsibility and reputation. The fifth level deals with self-actualisation of personal growth and fulfilment.

If the concepts of Schumacher's 'level of being' and Maslow's 'hierarchy' are incorporated into the framework of an organisation, we can divide organisations in four categories. A 'mineral organisation' caters to only the monetary needs of people. A 'plant organisation' fulfils monetary as well as security needs. An 'animal organisation' fulfils a person's monetary, security, social and ego needs. A 'human organisation' fulfils all the above needs, and as a bonus, people can aspire to accomplish their own potential. A human organisation motivates its people to give their best. Person is not a mere commodity in a human organisation.

A human organisation has 'designed' as well as 'emergent' structure. The designed structure, the formal structure, is created for a specific purpose. The emergent structure, or the social structure, is created by its informal networks and communities. Human organisations try to find the right balance between the 'creativity of emergence' and the 'stability of design'.

Human organisations, like living organisms, need inputs and feedback to grow and produce. They influence as well as are influenced by the environment they live in. They develop their own mechanisms for dealing with the stresses of environmental changes, and in the process learn the ways to insulate themselves from these stresses. Like an organism, they get sick if they can't cope with internal disorders and/or external pressures.

Competition is essential for human organisations. In the absence of competitive threats, they often fail. A human organisation, like a living organism, is selfish. It allows long-term changes to occur, only if there is productive advantage.

Human organisations have learnt from life sciences that the capacity to produce and diversify, and actual production and diversification are not the same. How much one can grow, depends on how much one can metabolise. An organisation must know how much it can really metabolise.

The hallmark of life is the capacity to associate, establish links, co-operate and maintain symbiotic relationships. A human organisation understands the need to shift from hierarchy to networking for organisational development.

As in human society, a human organisation respects the integrity of its organisational culture. The leader of a human organisation believes in a 'lean' organisation, but knows that 'dieting off fat is more difficult than not putting it'. A human organisation understands that the right to exist is not perpetual, but has to be continually earned. It recognises that growth will eventually stop if it has confused and impaired vision. An organisation dies when its sense of the self becomes meaningless.

Living organisms promote the philosophy of 'constructive destruction' - if something or someone has outlived its utility, destroy it. This is a hard moral question.

A human organisation is endowed with the power of reasoning, introspection, and communication. It is eager to learn and is flexible. It responds to new concepts and knowledge. It understands that the greatest risk is not to take a risk.

A human organisation tries to provide leadership like a good father gives to his children. He loves all his children, yet chooses the right one for a specific task. The father has genuine respect for his children, but doesn't hesitate to remove some from an assignment when it becomes clear that they can't perform the assigned task.

Continuance is the ultimate goal of any organisation. A human organisation understands that the right to exist is not perpetual, but has to be continually earned. It grows when it has the 'will' to grow and survive, and it sustains its growth when it begins to see the world beyond its own image.

5

Technology Revolution and Optimism

The way technology is evolving, and the way it is changing our ways of living, why should we not be optimistic about technology. But then optimism comes tagged with pessimism. We aim for unfathomable optimum optimism. Technology is like water. It finds its open spaces. If it fills the available spaces more than it should, there is flood.

Future engineers and technologists are expected to appreciate, more than before, the human dimensions of technology. They are expected to have a grasp of the global issues. There is a need to understand the nuances of working in a culturally diverse space. According to some experts, good engineering designs should not be deprived of the benefits of a broad spectrum of life experiences, as adequate familiarisation with societal demands is essential for practical technological literacy.

Future engineers need to be better equipped to deal with people of diverse backgrounds, such as from social sciences, management, and communication. Some of the grand challenges that are enlisted, from an engineer's point of view, are environment protection, hunger, energy and controlling the spread of the diseases. We need developers of responsible technologies and products.

We need managers to manage things, and at the same time we need adequate number of things to manage. There is a need to build faith in the public that engineers and technologists are sensitive to their concerns. One of the biggest responsibilities of the engineers and technologists is to keep themselves updated about professional developments and practices.

Countries, not only want exceptional scientists and engineers, but also people who are temperamentally innovators, and think they fit into their country's core values. Many would say, prepare young minds and nurture future innovators.

An optimist makes an important observation: Society is built on top of our infrastructure, not the other way round; "We built our school system on top of the industrial Revolution; we didn't get the Industrial Revolution because we'd sent people to school." And since our infrastructure has been stable for a very long time we expect similar kind of stability in our institutions that were built upon this infrastructure. He adds, that's about to change, because technology is expanding exponentially, and along with it is moving the infrastructure development.

To accept any major change, society needs 'social and organisational innovations'. Infrastructure is becoming more and more digital. The development costs are coming down. 'Stability' is assuming new meaning; "the thing you did to get there will no longer work to keep you there." One may therefore ask - Is the society ready to accept such exponential development?

Besides technological innovations, it is important how we prepare our institutions, infrastructure, and society to accept the changes. It is important to make our institutions networked, distributed, yet able to maintain their identities. Foremost, we need good leadership to run the institutions and build networks and infrastructure.

'Too much is changing too fast' is the sign of the time. Products decide the tool boxes. Education is no exception. The meaning of education is changing very fast. It is difficult to predict what you will need to know after 5 years. Long term plans are passé. No one knows what an imaginative kid would like to become when she/he grows up. It is an evolutionary process, and, often, evolution is not a step-by-step process. A 'factory' model of education is not fit for the present time.

Future is becoming beyond the realm of the 'astrologers'. One only wishes, not to get 'the future shock' as big as 'the big bang'. If I have to choose between optimism and pessimism I would go for optimum optimism.

Rational optimism holds. "There is a great deal of human life that does not change," Matt Ridley (1) pointed out. To support his argument he says, "Natural selection is a conservative force. It spends more of its time keeping species the same than changing them."

We, howsoever, would like to retain our individual identity. It is difficult to survive without the aid of collective intelligence. Humans are humans, because "at some point in human history, ideas began to meet and mate, to have sex with each other." Ridley adds, "Evolution can happen without sex; but it is far, far slower."

Science, technology, and engineering are so interlinked that one can't be separated from the other. Technologies have significantly altered the way we have lived, and we will live in the future. Science will not end, engineering will take new forms and technology will give new products. We do not know much about 'yet-undreamed-of disciplines' that will give us unimaginable products.

History tells us that every scientific theory has a half-life; a new theory replaces the old one. The social endeavour will keep on pushing scientific endeavour. Science will not die, until the day scientists will remember their social responsibility. And we are very hopeful that scientists will not forget their responsibilities. If they forget it is their doomsday.

Science will continue to fly in different horizons; its mode of transport will be different, its ultimate destination will not change, and that will be to serve the society, from the viewpoint of humanity at large.

Genes have not changed much, but the society has, with time; primarily due to collective intelligence and enhanced diversity. "The more individuals it could support, the more habits it could acquire. The more habits it acquired, the more niches it could create." One of its consequences is overpopulation.

In an overpopulated world one is more for oneself. When one is more for oneself, consumption increases; wants become more than needs. The situation is like, as Ridley pointed out, species expanding in number but not in living standards. To get over this imbroglio species will devise new ways to fill in the bag of collective intelligence, and that will help the growth to continue.

Another kind of problem the future world will face will depend upon the kind of world it wants to inhabit. The world will never be free of parasites or predators. Both parasites and symbionts will exist, as they should, for the existence of each other. The future knows how to keep the balance between predators and preys. The pool of collective intelligence will continue to get filled, in spite of the presence of parasites and predators. The innovation tap will never be dry.

The future world will be the world of individuals; a decentralised world for the collectives; one will feed the ego of the other for their own survival.

It is true that history can't be the true guide of our future, but history can help us in optimizing the optimism of the progress. History recognises that all progresses are not desirable, nor sustainable. Past may not predict the future, but without it there is no future. The big asset of the future is its past. History can tell us so much about poverty, hunger and illness. Optimum optimism is a more realistic attitude than 'apocalyptic pessimism'. I would wish that the Government has little say in the future world in the matters of organizing it. I wish the impact of the future shocks will be beneficial to the mankind, as was the impact of the big bang.

Every moment we are changing, as a result of engineering interventions and innovations. We are producing huge amount of data. Processing of large volume of information is one of our big problems. We are developing digital data management protocols that are secure, transparent and user friendly for various kinds of transactions to rid us of intermediaries.

Artificial intelligence (AI) is our big bread-and-butter source as well as headache. Perceptual learning, memory organization, and critical reasoning are not easy to implement. But nothing is difficult, if there is market. We all love to anticipate financial opportunities, what better if AI does it. If robots can share some of our tasks, why not let them do it. Why should we object if they 'willingly' take care some of our maintenance responsibilities. Self-driving cars are anticipated to solve many problems (such as accident, pollution, noise) , as well as create many problems in the name of progress. We are entering the world of virtual and augmented reality in a big way to know our environment somewhat better. The 3D printing revolution will allow us to create products in all possible forms, quickly and accurately.

Using emerging technologies, it seems, it is possible to do anything and produce anything. We can turn back the clock, we can grow pesticide free organic food using less water and energy in 'Vertical Pink farms'; atmospheric water harvesting (collecting water and moisture from the atmosphere) can be an alternative to drilling water; human enhancement, including wearables and implantable, seem a distinct possibility.

In the list of the most intriguing inventions of Technology Review of 2018 are included: (1) Anti-aging medicines (boosting elderly peoples immune system, eliminating senescent cells that make aging bodies break down), (2) Electric planes with no moving parts (using electro-aerodynamic propulsion to create íonic wind' to push the plane forward), (3) DNA computing for programmable pills (that detect the right signal to target infections), (4) Group brain-to-brain communication (that transmits thoughts of one to the other), (5) Tracking people on the other side of the wall using Wi-Fi (detecting movement distortion signals from any WIFI transmitter in the area), (6) Quantum communications via satellite (using quantum cryptography), (7) Phones that shoot a million frames per second, (8) Edible electronics (disposable electronic circuits to track the effect of drugs on health), (9) Electricity-generating boots to power small communication devices.

In any democratic set up it is important to know what people know, want, believe, and fear about. The views of the people change, as the science and technology changes. In this context, ask Jennifer Hochschild, Alex Crabill, and Maya Sen (2) - Is knowledge about genetics associated with more optimism about genomic science? Another question they ask is - Do religiosity or characteristics such as race or gender play a role in levels of optimism about genomics in general or particular genomics arenas?

They make an important observation: “citizens differ from most social scientists, legal scholars, and policy advocates in their overall embrace of genomics’ possibilities for benefitting society.” They say, “Technology optimism is the ‘underestimation and neglect of uncertainty’ in favour of ‘widely shared speculative promise’. Technology pessimism, on the other hand, “is the overestimation of threat and harmful impact and insufficient attention to benefits or to people’s ability to respond appropriately to risk.”

A pessimist is more concerned about security risks and possible losses, even if that means missing opportunities of potential gains. Commitments are often made ‘behind closed doors with insufficient public scrutiny’. The lack of relevant information and analytical framework among the common people results in contradictory reports from the media people and opinion leaders; “psychological proclivities shape reception of messages.”

Some perceive unjustified optimism over a wide variety of tasks. There is always a chance of mismatch between what one would like to see happen and what is objectively likely. Some perceive widespread pessimism; they perceive that they face more risk now than in the past, and they will face greater risk in the future.

“Education and scientific knowledge affect perceptions of risk, as do social influences and communication with members of their social networks.”

Technology Wants What Life Wants

Technology wants what life wants. Life wants increasing efficiency, increasing opportunity, increasing emergence, increasing complexity, increasing diversity, increasing specialization, increasing freedom, increasing mutualism, increasing beauty, increasing structure and increasing evolvability.

Technology is facing the dilemma of want and need. Need is limited, want is unlimited. Need is a necessity; it defines the limits of enough. Want is an optional; it defines the depth of appetite. We don’t know where need ends and want begins.

Technology suffers from the constant tension between the ‘virtues of more’ and the ‘necessity of less’.

Kevin Kelly (3) coins a term ‘Technium’. The term includes, besides hardware, culture, art, social institutions and intellectual creations of all types. It includes intangibles like software, law, and philosophical concepts. It includes the generative impulses of our inventions to encourage more tool making, more technology invention, and more self-enhancing connections.

"The technium accelerates the invention of technologies." Kelly says that technium is maturing into its own thing. "It may have once been as simple as an old computer program, merely parroting what we told it, but now it is more like a very complex organism that often follows its own urges."

Technium is as Complex as Mind is.

We say that an entity is autonomous, if it displays any of these traits: self-repair, self-defence, self-maintenance (securing energy, disposing of waste), self-control of goals, self-improvement. The common element in all these characteristics is of course the emergence, at some level, of a self. Any system doesn't display all these traits; some systems display some of them.

In addition to these drives, technium has its own wants. It wants to sort itself out, to self-assemble into hierarchical levels, just as most large, deeply interconnected systems do. The technium also wants what every living system wants - to perpetuate itself, to keep itself going. And as it grows, those inherent wants gain in complexity and force.

More than half of the living species on this planet are parasitic. Biologists believe that every organism alive (including parasites) hosts at least one parasite. This makes the natural world a hotbed of shared existence. As life evolves, it becomes increasingly dependent on other life. As life evolves, nature creates more opportunities for dependencies between species. Every organism that creates a successful niche for itself, also creates potential niches for other species. As life evolves, possibilities for cooperation between members of the same species increase.

Technology is evolving, because the human mind is evolving. The evolvability of technology depends upon human minds evolvability. The human mind will decide the limits of technology. More the human mind expands the space of possibilities, more will technology grow.

"Increase your options but follow the principle of minimalism to make living optimum." In other words, learn to seek the minimum amount of technology that will create the maximum of choices.

There are two kinds of games: finite and infinite. A finite game is played to win, and the game ends when someone wins. An infinite game is played to keep the game going. It does not terminate, because there is no winner. In finite games, rules remain constant. If the rules change during the game, it is termed unfair. An infinite game, however, can keep on going only by changing its rules. To maintain open-endedness, the game must play with its rules.

Technium, like mind, is infinite game. Its goal is to keep playing—to explore every way to play the game. One technology can have different applications; what one designs, perhaps, has a moral purpose.

Writes Jason Pontin (4), "We are technology-making apes who evolve through our material culture; everywhere, people fly like birds, speed like cheetahs, and live as long as lobsters, but only because of our technologies."

The three commandments, as proposed by Pontin for technologists are:

(1) Design technologies to swell happiness. Its corollary is: Do not create technologies that might increase suffering and oppression, unless you're very sure the technology will be properly regulated.

(2) In regulating new technologies, balance costs and benefits, and work with your fellow citizens, your nation's lawmakers, and the world's diplomats to enact reasonable laws that limit the potential damage of a new technology.

(3) The best technologies have utility, but also provide fresh scientific insights. Prioritize those. If there is a need to develop a technology, it is developed to meet that need. To develop a technology suitable environment is required. There is always a possibility that one environment is better than the other to develop that particular technology. It simply means, developing a particular technology requires several inputs.

Jared Diamond (5), in his book Guns, Germs, and Steel, tries to answer - Why some regions are different than other regions? What factors cause the gap between the development of one culture and another? Diamond says, some societies are more materially successful than others due to their geographical location, immunity to germs, food production, the domestication of animals, and use of steel. His book gives an account of more than 13,000 years of human evolution and societal development.

Mapping the history of migration, Diamond concludes that it is geography, not biology or race that produced cultural disparities. Those populations thrived that moved from hunting and gathering to cultivating crops and raising domestic animals. Farming societies eased the burden of producing food and are then able to devote their time to making weapons and perfecting the art of using them. The successful regions had a natural advantage in agriculture because of the presence of plants and animals that were easily domesticated. Empires with steel weapons were able to conquer or exterminate tribes with weapons of stone and wood.

Diamond writes, “Blessings of civilization are mixed.” Some industrialized states may enjoy better medical care and a longer life span, but may also receive much less social support from friendships and extended families. His motive, he writes, is not to celebrate one type of society over another, but simply to understand what happened in history.

“Geography can account for history’s broad pattern,” is Diamond’s submission. His point is – racist explanations are not going to resolve these questions. “From the very beginning of my work with New Guineans, they impressed me as being on the average more intelligent, more alert, more expressive, and more interested in things and people around them than the average European or American is,” writes Diamond.

In spite of superior intelligence, why then the New Guineans remained technologically primitive? Has climate something to do with it? “Perhaps the seasonally variable climate at high latitudes poses more diverse challenges than does a seasonally constant tropical climate. Perhaps cold climates require one to be more technologically inventive.”

The main theme of Diamond’s book is: “History followed different courses for different peoples because of differences among peoples’ environments, not because of biological differences among peoples themselves.”

Technology is known for efficiency, utility and productivity. It offers competition, exhilaration as well as stress. It closes the gaps, as well as creates. It liberates, as well as enslaves. It brings people closer, as well as creates deep divisions. It brings to the fore our endless desires. It is growing endlessly on this finite earth.

“Moore’s law, according to which the speed of computer chips doubles every two years, now seems to apply to life itself.” One of its consequences is that we have begun to regards many useful things as ‘impractical luxury’. We can’t see the beauty in slow moving butterflies. We have time to barely register an image, and no time to actually see it. We are losing our ability to “endure the long shot, the slow dissolve, the sustained monologue.” We are opting for faster version of life, but we are not ready to handle it. Our vision is in transition. We come across greater number of visions, but lesser number of experiences. We have photographic memories, but no time to enliven the memories. Amidst plenty we are missing subtlety.

We live in a less violent and more powerful world, yet we are unhappy. Designed happiness in a consumerist society looks manipulated. We are willingly paying the price for our more comfortable living. At the same time we do not wish to be the slaves of technology. We do not wish to be spiritually hollow. We do not wish to die enclosed in a PC. We do not wish to receive wishes from the departed souls from the sky. We do not want to live beyond the threshold in an unnatural world. We want to live in a natural world.

No one knows where our future lies, still we make projections. The visionaries among us are both utopian and dystopian.

Utopian visionaries see in the future world a state of balance and peace, and where all life is valued and sustained. Since the world has achieved its full potential, they see no reason for one to be aggressive. They maintain that man has no enmity or competition with nature. In the future world they find no difficulty to expect a world of equal opportunity and equitable distribution of goods and services. They see in the future world abolition of cultural, racial and gender-based prejudices. They believe that humanity has solved all its problems with the help of sensibly developed and rightly used technologies.

Dystopian visionaries, on the other hand, imagine a future world where life and nature are recklessly exploited, and eventually destroyed. They predict catastrophic destruction of our natural environment. They imagine the loss of complete freedom of the mind due to technological interventions. They believe future generations will depend more upon artificial intelligence than their native intelligence. They believe technology will make them slaves of technology.

Utopian future is projected by the idealist visionaries. Dystopian future is projected by those who feel oppressed by their environment, and are afraid to fight extreme odds. Overly disastrous future projection is definitely not a great idea, as excessive optimism is. The point is to avoid utopian and dystopian extremes and take a conciliatory middle path. If optimism is mixed with some amount of pessimism, it works better. If in certainty, a certain amount of uncertainty is mixed, it generally leads to better end results.

In the techno- bulged society it is time to make real choices using smart machines, and for that is required human intervention and inner transformation.

What For Education Is?

What are the Universities for? Are they to prepare graduates for work or to make good citizens? Both are needed; understanding human relationships is as essential as understanding the technology.

Has education changed over the years? Is the change reflecting changes in learning outcomes, particularly in social and global awareness, ethics and professionalism, group skills, and application of engineering skills? Is the out-of-class experience influencing student learning? Has the learning improved problem solving ability? How do we plan for something (like the learning process) we can't predict?

Who is responsible for the educational ineffectiveness in our schools? Is it due to the inadequacy in the teachers, or absence of market incentives, or inequities of societies, or lack of commitment to excellence?

It is believed that outside the classroom model deepen knowledge. It is also said that traditional lecture format has stood the test of time because it is effective. One may argue in favour of the new, but also agrees that it is difficult to dislodge the old. Moreover, many are not fully convinced that the new method is fully useful. According to one study "disciplinary differences were important to take into consideration – engineering was particularly well suited to flipping." The study suggested that "blended learning appear to have a stronger effect on student learning outcomes".

The development of cross-disciplinary academic structures; for example, a course in Manufacturing, as in Cambridge University, could be a collection of expertise in management, engineering, technology and policy surrounding manufacturing.

Do smart classrooms serve their purpose? Some educators doubt the utility of excessive use of technology in schools. They doubt the wondrous power of technology to "enable, motivate and inspire" students. Diana Schaub of University of Maryland thinks, "What the technology activated was their narcissism and vanity. They were intrigued by the instantaneous information about themselves. They gloried in the 'knowledge' of how many of their fellows deserved to be mocked for their ignorance." She thinks drawing students out of their narrow self-absorption doesn't help students. The emphasis on computer-assisted engagement threatens to short-circuit the connection between moral and intellectual virtue. "When listening intently, the self is humbled or silenced, while at the same time becoming fuller and more expansive".

Researchers, however, have not been able to establish a clear link between computer-inspired engagement and learning. They say smart classrooms don't always churn out smart students. One needs teachers who can earn the attention and respect of their students.

We need both 'emotionally engaged' students as well as teachers. Deborah Stipek of the Stanford University School of Education suggests some ways of improving emotional engagement. Her suggestions include: connectivity of the subject matter to students' personal lives and interests, opportunities for solving novel and multi-dimensional problems. It is true that technology promises empowerment, but it should not be allowed rival or displace the essential relationship of teacher and student. We know that competition takes away the joy of learning. Technology should not become a means of aggravating competitive pressure among the students.

Though education is the vehicle of social mobility, many educators believe you don't generate growth through number of graduates. Quality matters much more than quantity, not only in education, but in most spheres of human activities. Alison Wolf (6) thinks quantity at the expense of quality is counterproductive. She says if two aspirins are good, it doesn't mean five aspirins are better. She grew up with the easy and affordable teaching tool, chalk and blackboard. The chalks are now dust free and the boards are made of better material to make writing and rubbing better, and therefore there should not be any possible continuing with these tools.

We like to accept new things, and at the same time, we like to discard old things. We prefer, compared to traditional class rooms, flipped classrooms. A recent article discusses the pros and cons of both the methods. Introduction of a new system is always full of promise. The pitfalls in new system too are many.

How higher education will look in the future? Will intelligent machines usurp the jobs of teachers? Will the concept of individual campuses slowly disappear? Will the exams that emphasise mastery of taught knowledge no longer be the primary tool for judging student's performance?

Our memories have great flexibility and creativity. We like to understand things by entering into other's minds through their eyes and ears. And as Oliver Sacks writes, "Memory is dialogic and arises not only from direct experience but from the intercourse of many minds".

Engineering and Human Progress

Engineer is a necessary link in the chain of human progress, wrote W F Durand (7) in 1925. Engineer develops and translates into use the needs of civilization. While doing such function, he acquires the weighty duties and responsibilities of the society. In the same vein, Dugald C Jackson (8) tells us that engineering is not only spectacular physical entities. The real engineering is an "intellectual phenomenon out of which these and infinitude of other physical symbols come to useful birth. This true engineering is a process rather than a physical result – the process of planning, organizing and executing works intended for identifying and directly the forces of nature to man's service." An idea is generated in the mind. We need tools to link the products of our mind to take them further in the value chain. Is science driven by ideas or tools, asks Freeman Dyson (9). For Dyson, tool-creation has been indispensable to scientific progress. "Philosophy supplied the concepts for science, and skilled crafts provided the tools." Dyson refutes the idea that says that scientific revolutions are concept-driven. Dyson argues, "We remain tool making animals, and science will continue to exercise the creativity programmed into our genes." Discovery is not enough to account for change. His famous statement is: "In every human culture, the hand and the brain work together to create the style that makes a civilization."

We're advancing technologically, but why is the world not producing scientific geniuses as it used to? Novelty is generated randomly or strategically? Is creative thought a process of honing or selecting? These are important questions to understand creativity and discovery.

Dean Keith Simonton (10) thinks the world has stopped producing geniuses. He thinks, we may not have another Darwin or Einstein. He, however, believes that Darwin and Einstein are needed, as there are still many fundamental questions to resolve. For a genius to emerge, some think, the world must fall into some kind of crisis. The good thing is that the world will always have a few unresolved crises, and thus we should always hope for the emergence of geniuses to resolve them. Denis Dutton writes about Picasso —in order to overcome "an indigestion of greenness", Picasso walked into the woods to "empty this sensation into a picture."

I have understood that confident engineers are expected to foresee and manage the unknown and unexpected problems. They are expected to appreciate, more than before, the human dimensions of emerging technologies. They are expected to understand global issues, and the nuances of working in a culturally diverse space. I have understood that a fairly good grasp on inner engineering is required to do any kind of good engineering.

We need different kinds of engineering educators. We need to shift the focus of engineering curricula "from transmission of content to development of skills that support engineering thinking and professional judgment." The challenge of future engineering education is to weave its various components — engineering profession, the society it serves, educational institutions, educators, and students — into a whole.

Engineers need to ask themselves the most important question — do I take pride in designing a thing and manufacturing it, as I take pride in packaging it.

Engineering Ethics

Ethics is an important aspect of any profession. It is good to remember that, as someone rightly pointed out, "The teachings of Plato, Mill, Kant, Spinoza, Descartes, Nietzsche, Epicurus, Confucius, and others will indeed provide a very solid foundation for the understanding of ethics. But it is important that ethics courses also deal with the pragmatic issues that confront engineers in the rough-and-tumble, everyday world in which they live and work."

Future engineers are expected to appreciate, more than before, the human dimensions of technology. They need to understand the nuances of working in a culturally diverse space. On the one hand, we are not limited by technology, and on the other, we are worried about risks to the environment, health and safety. In this changing scenario, competition is natural. But, the new world also wants methods that are appropriate to make cooperation natural.

We can't live in an isolated world. We are witnessing organisations benefiting from alliances and joint ventures with competitors. We are seeing the virtues of 'cooperative competition'. According to some experts, the merger of creative disciplines (art, music, architecture) with engineering activities (design and innovation) will add value to the engineering profession.

Good engineering design should not be deprived of the benefits of a broad spectrum of life experiences, as adequate familiarisation with societal demands is essential for practical technological literacy.

Engineers need to change the public image of engineering. They should not remain 'behind the scene', as they have traditionally remained. They should be present wherever they matter.

Are the minds of engineers different? Is it a combination of opportunity and resources? Are we cultivating the right kind of engineering mindset? Do only the academic grades reflect the quality of an engineer? What is more important for an engineer – insight or precision? Can a person trained to solve expected problems able to deal with unexpected problems? What a general engineering toolkit must contain? Is there a need for various specialized engineering streams at the undergraduate level? What additional efforts are required to impart practice- based experiential knowledge? Shouldn't the practice of industry mentoring be taken more seriously? Why most engineers don't take as much pride in designing a thing and manufacturing it, as they take pride in packaging it and marketing it? Are there enough challenging jobs in the industry to attract good engineers?

Developments in different areas of science and technology are adding new dimensions to the ethical issues. A glimpse of future ethical issues can be had from UNU Millennium Project. Since "some future issues are further in the future than others," UNU Millenium Project grouped the questions related to ethical issues into three time periods: 2005-2010; 2010-2025; and 2025-2050.

Some of the relevant issues pertaining the past (2005-2010) were: Is it right for governments or the public to intervene in the scientific process when, on the one hand, unimpeded science has such great promise but on the other, unintended deleterious consequences are a plausible result of the research? Do people and organisations have a right to pollute if they can pay for it; for example, by paying carbon taxes, pollution fines, carbon trading, etc.?

The issues pertaining to the present (2010-2025) include: Should there be two standards for athletic, musical, and other forms of competition: one for the un-augmented and another for those whose performance has been enhanced by drugs, bionics, genetic engineering, and/or nanobots? With a vastly more interconnected world, when ideas, people, and resources can clearly come together to solve a problem or achieve an opportunity, is it unethical to do nothing to connect them, when it is clearly in one's power to do so?

The issues pertaining to the future of ethics (2025-2050) relevant to science and technology include: If technology grows a mind of its own, what ethical obligations do we have for its behaviour? Do we have the right to genetically change ourselves into a new or several new species? Is it right to allow the creation of future elites who have augmented themselves with artificial intelligence and genetic engineering, without inventing a way to manage their superhuman abilities?

The Inner Engineering

More and more, we are getting drawn into the world of data. We are becoming electronic and biochemical algorithms. Data are empowering us. A day will come when we will use the same basic tools and the 'common language' to understand entirely different disciplines represented by different set of data. We the humans transform data into information into knowledge into wisdom. When there is Big Data, it is humanly not possible to process it using the brain, and in such situations there is possibility of the chain to break, and if we don't want to break the chain, we need to process the data through electronic algorithms. Organisms are algorithms; they are seen as data processing systems.

Writes Yuval Noah Harari (11) "As both the volume and speed of data increase, venerable institutions like elections, political parties and parliaments might become obsolete – not because they are unethical, but because they can't process data efficiently enough. These institutions evolved in an era when politics moved faster than technology. Though the modus operandi of politics has not changed much, technology has shifted its gears from first to fourth. Technology is likely to outpace politics completely in the near future. "The governmental tortoises cannot keep up with the technological hare." As a result, power will be shifting from the politicians. But no one yet knows where it will shift to.

It is said that the world is not producing visionaries the way it used to in the twentieth century. Our present day visionaries are not in a hurry to build a new world, unlike their predecessors, though they have better technological means to do so. As Harari writes, twenty-first century politics is bereft of grand visions. "Government has become more administration. It manages the country, but no longer leads it." Harari adds, "Given that some of the big political visions of the twentieth century led us to Auschwitz, Hiroshima ...., may be we are better off in the hands of petty-minded bureaucrats. Mixing godlike technology with megalomaniacal politics is a recipe for disaster." For the politicians, profound wisdom is inaction and ignorance. A combination of godlike technology and myopic politics can also lead to disaster. So we have to be careful to deal with this situation. We can't leave ourselves at the mercy of market forces. We have to find out what is good for us. We have to find out that 'somebody' who will be in charge. Harari's most interesting prediction: once the most efficient data-processing system, Internet-Of-All-Things, is in place, "Homo sapiens will vanish."

In spite of so much 'progress' we are worried about our future. Robert Pollack (12) makes an important observation: "we threaten the planet by our success." Pollack's concern is our success, not our failure. For Pollack success is "survival of the future". Success does not mean only winning, it also means conservation of the future. "I'm interested in the same old boring thing inside a mortal universe of mortal people—how best to care for each other and to care for each other's futures."

After I had almost completed this book, the book Inner Engineering by Sadhguru (13) I ordered a few days ago, arrived. While glancing through the first few pages of the book, I felt joy in my ambience. It was a good feeling.

The book at one place says, "We have tremendous tools of science and technology at our disposal...However, if the ability to wield such powerful instruments is not accompanied by a deep sense of compassion, inclusiveness, balance, and maturity, we would be on the brink of a global disaster." At another place it says, "Some are suffering their failure, but ironically, many are suffering the consequences of their success. Some are suffering their limitations, but many are their freedom....Everything is in place, but the human being is not in place....You cannot transform the world without transforming the individual.... Your joy, your misery, your love, your agony, your bliss, lies in your hands." This is 'inner engineering'. It is soft engineering.

"Inner Engineering is an opportunity to engineer an inner transformation and deepen your perception, bringing about a dimensional shift in the very way you look at your life, work and the world."

Sadhguru writes, "So truth is timeless, but the technology and the language are always contemporary...So, while I will be exploring an ancient technology in this book, it is also a technology that is flawlessly state-of-the-art."

Everything evolves. Life evolves. Mind evolves. Education evolves. Science evolves. Engineering evolves. Technology evolves. Biologists believe that every organism alive (including parasites) hosts at least one parasite. This makes the natural world a hotbed of shared existence. Taking a cue from the natural world, it can be said that in the interconnected world everything evolves with the support of everything else. If life evolves, mind evolves. As mind evolves, science evolves. As science evolves, engineering evolves. As engineering evolves, technology evolves. There is another way to look at it. It says, for example, technology evolves when its time comes. What technology wants it gets. It is said, "Technology creates itself out of itself".

Inventions are one endless chain of parallel instances. It doesn't leapfrog. It takes its own time to develop. It proceeds, like evolution, to the 'adjacent possible'. In many senses, technology qualifies as a living organism; it is self-organising, can reproduce, and respond and adapt to its environment. Moreover, it follows its own urges. As Matt Ridley writes in The Evolution of Everything (14), "The implications of this new way of seeing technology, as an autonomous, evolving entity that continues to progress whoever in charge, are startling. People are pawns in a process. We ride rather than drive the innovation wave. Technology will find its inventor, rather than vice-versa."

Some say freewill is an illusion. I would like to believe that freewill is not an illusion. As pointed by a biologist, actions simply reflect the genetics of the organism and environmental history; genetics and environmental history of us only. Some say, we have inherited a belief in free will. It is difficult to accept that we are nothing but the neural signals of our brains, howsoever scientifically authentic it may look. Freedom evolves. It evolves to different extents. The freedom of the bird is qualitatively different than the feedom of the humans. Free will is not 'all-or-nothing', points out Daniel Dennet. "Technology, science, knowledge, human rights, the weather forecast – they all increase your freedom to alter your fate," writes Ridley. It is safe to assume "that the soul and the will are real things with no history and no trace of their origin."

The world of engineering is changing. A new mind is evolving to deal with this world. Engineering ethics is changing as well. Some old ethical questions shall remain, some old issues will be forgotten, some new questions will be asked. As the UN's Millennium Project says, one of the future issues of engineering ethics is, if technology grows a mind of its own, what ethical obligations we have for its behaviour.

In the knowledge age both the 'tiny systems' and 'macro systems' are playing a central role. The inputs needed to make a system functional are theories, tools, and resources. Science gives theories. Technology gives tools. Nature and society give resources. Science and technology give theories and tools based on the inputs it receives from the nature and society. Engineering in return gives back to the society and nature products and benefits. Each node of the network – science, technology, engineering, nature, society – is intimately connected to the other. A weak node disrupts the whole system.

There are reasons to be optimistic about the future. There are also reasons to be worried about the future. Only the fittest survived in the past. Only the fittest will survive in the future. There was competition. There will be competition. The question is - What does survival of the fittest would mean in the future? The conventional view of the fittest is the one who is aggressive, violent and selfish. But it also means loving, least aggressive and selfish.

Is technology changing the nature of moral emotions?

A new understanding of evolution says that the driving force of evolution is not found in the chance events of random mutations, but in life's inherent tendency to create novelty, in the spontaneous emergence of increasing complexity and order. This view has taken our attention to the importance of cooperation in the evolutionary process. Many believe, life did not take over the world by combat, but by networking.

We alter our environment to meet our needs. Our survival is based on adaptability, connectivity, communication and cooperation. If we want to evolve the whole ecosystem has to evolve. If I am vulnerable, I will like to protect myself. I would like to become stronger and stable. I would like to prepare myself to face competition. I would prepare myself for competitive cooperation. Cooperation is as essential as competition for continuation. A selfish doesn't like another selfish. "The best groups might be those that include a few devils along with the angels."

We are not blindly programmed robot vehicles. We understand that evolution is also manifestation of consciousness. Every individual is unique, and our uniqueness helps us to survive in the demanding environments. But we need exercise care and restraint. As individuals it is our responsibility to care for the welfare of others.

6

Future of Work

"I think that there is far too much work done in the world," writes Bertrand Russell in In Praise of Idleness. Rightly so, a workaholic is not necessarily a valued employee. Why we work, asks Barry Schwartz. Why for most of us work is monotonous and meaningless? Why we have twice as many "actively disengaged" workers in the world as there are "engaged" workers?

Do we work only for money? Some think, work is about pay and nothing more. It is true, people work for money, but not always. Many people work for the challenge, meaning, and engagement it brings. "The lesson here is that just how important material incentives are to people will depend on how the human workplace is structured. And if we structure it in keeping with the false idea that people work only for pay, we'll create workplaces that make this false idea true," writes Schwartz.

Daniel Kahneman (1) thinks, education is an important determinant of income, but not as important most people think. While heavily focussing on education, thinks Kahneman, one tends to miss many other things that determine income. This mismatch is the cause of 'focusing illusion'. No doubt education is important, but often we don't know what we want from it. Kahneman observes: "Our large commitment to there being good schools ironically has not been matched by concern about what they are for."

Mihaly Csikszentmihalyi's (2) concern is the aim of education. He found lack of alignment between the educators and the students, concerning the aims of education. Teacher's responsibility is to provide cognitive tools, like critical thinking and analytical skills. On the other hand, the priority of the student is to learn how to live a good life. These two objectives are not necessarily contradictory, often they are. Cognitive tools are acceptable, only if they contribute to good life. Often cognitive tools are considered by the students "as a tiresome demand unrelated to their true needs." Teachers think of happiness "as a hedonistic yearning unrelated to the pursuit of knowledge."

What is good work? Good work is what gives people "meaning, engagement, discretion, autonomy, and opportunities to learn and grow." Good work for

Howard Gardner (3) is the integration of excellence, ethics and engagement, empathy, and equity.

For Susan Verducci (4), both trust and accountability are important catalysts of good work. Her argument is "when accountability becomes a focus, trust diminishes, and when trust takes centre stage, the need for accountability recedes." Moreover, trust and accountability are both responses to uncertainty and risk. We trust, in spite of the fact that it involves risks. We trust, because it evokes positive actions and feelings in others. "Trust lowers transaction costs by not requiring costly activities such as monitoring others' behaviour." It essentially 'lubricates' interaction and cooperation, while the consequence of its diminishment or absence is friction." ((5).

Accountability also responds to uncertainty and risk in professional work. It is expected to meet the agreed upon expectations. It can lessen, if properly conceived and implemented, the unexpected or unwanted consequences. Accountability raises awareness on critical issues, gauges what is working and not working, opens doors of communication, allows for midcourse changes in strategy, and can make work more transparent. But there is a glitch. When we want the benefits of accountability, we may have to sacrifice the benefits of trust. When we prioritize the benefits of trust, we may incur the costs of the absence of accountability.

Workplaces are important, as they are the important drivers to shape human nature. If we design workplaces that permit people to do the work they value, we will be designing a human nature that values work. People want to work if the work they do is of value to them.

The role of learning for the development of a workplace has been emphasized by Peter M Senge (6). In a learning organization people continually discover realities, and learn how they can be changed. But, as Senge pointed out, most organizations have poor learning abilities. Senge identifies a few learning disabilities:

1. "How could I do anything else" mentality. It is not easy to get over this identity crisis. It is good to remember that there exists a 'complimentary world' beyond our 'own little world'.
2. "There is someone or something outside ourselves to blame when things go wrong". There is a saying that if you see smile on the face of a politician, you know that he has identified a person whom he would blame for his own follies.
3. Too much proactiveness often is reactiveness in disguise. We want to solve problems before they become crisis. Sometimes this approach misfires due to aggressive actions.

4. "Event fixation"; the belief that for every event there is one obvious cause. We often forget that the threat to our survival comes not from sudden events, but from slow gradual processes.
5. Not noticing changes until they become threats, and then there is not much to do. We are geared to tackle sudden changes, and not slow gradual changes. We need to learn to pay attention to subtle as well as dramatic.
6. "Delusion of learning from experience"; We learn best from experience. When our actions have consequences which are indirect or happen in the distant future, experiential learning becomes difficult.
7. In order to keep the cohesive image of the organization, the management makes unjustified compromises. This 'skilled incompetence' keeps away the management team from learning.

Though virtually any job has the potential to offer people satisfaction, many people hate to do what they do. Jeffrey Pfeffer's book The Human Equation (7) elaborates on this subject. The effective organisations, according to Pfeffer, provide a high degree of employment security (including better pay; individual incentive is not encouraged), which builds employee loyalty and trust. In such organisations, employees are given a lot of discretion and autonomy, and are also provided extensive training. Such organisations rely on decentralized decision-making and all employees benefit through some form of gain sharing. Over-measuring performance of the employees, Pfeffer believes, does not yield desirable results. He observes, when a company earns low profits due to higher costs and poor customer service, the immediate tendency of the company is to cut costs. This means reduction in salary and layoffs, freeze on hiring. Training cost is also reduced. These result in decreased worker motivation, worse customer service, less job satisfaction and more turnover. Obviously, more trouble to the organisation. In other words, "cure makes the disease even worse." This 'vicious cycle' can be broken by introducing 'virtuous cycle'. The component of the virtuous cycle is to once again finding engagement and meaning in the work. How to make them happy to work? Barbara Fredrickson (8) suggests a few measures. She says, when people are in a state of positive emotion they think expansively and creatively. Positivity nurtures an environment of positivity. Negativity shows us the ways of taking wrong paths. Positivity creates a win-win situation for all.

The realities of a workplace can be quite different from its stated policies, or public face. In order to keep the cohesive image of the organization, the management makes unjustified compromises, hush up disagreements, etc. Decisions are taken at the administrative level, but input is rarely collected from the personnel expected to carry out the changes.

Organizations want to see a return on any investment. Employees on the other hand, want to see direct benefits to themselves. Openness to new ideas may tend to promote actions contrary to organizations vision or values. Workplace reorganization may not always value or empower workers. Some could be laid off, in the interest of the organization. We must decide how much we can compromise, and at what point we might need to withdraw. Job perspective of new employees (more mobile, willing to sacrifice job security for new challenges, better financial incentives, improved work conditions, flexibility in work schedule, location, etc.) are different and this may lead to potential conflict with old employees. Surely, we can't leave organisations solely in the hands of accountants. They are afraid of the 'red' colour. They have to understand 'red' and 'green' on the basis of broadened definition of inputs, and that includes 'psychic' costs, and the benefits organisations accrue.

Leadership is a dynamic phenomenon. It plays an important role for the growth of an organization. Leadership, built over time, has good nose for talent, and possesses low tolerance for nonsense. A true leader has a global genome and an innovative mindset.

Leadership, initially was to manage the concerns and outcomes of smaller groups. It then shifted its attention to organizations. Leadership is both 'individual' and 'collective'. It weighs 'threats', as well as 'opportunities'. It evaluates strengths, as well as weaknesses. Equipped with clarity of vision, it wants to see the big picture. Leader is expected to have the ability to communicate messages clearly. Leader is expected to face difficult and ambiguous conditions, and therefore, is expected to have envisioning and energising leadership. At the one end, leader is required to satisfy workers, and at other end, leader has to sing the songs that board likes.

Warren Bennis (9) says that the most reliable indicator and predictor of 'true leadership' is an individual's ability to find meaning in negative situations, and to learn from the trying circumstances. Universality of leadership is a widely debated idea. Essential attributes of leadership vary. It varies, depending upon the organisational structural and cultural differences.

Some cultures have 'prototype' expectations from their leaders. Abraham Zaleznick (10) points out that managers and leaders are different kinds of people; they differ in motivation, personal history, and in how they think and act. Leaders build and shape ideas. Managers give solutions. The leadership envisaged by Jim Collins (11) attains different levels: 'Effective Leader' is expected to catalyse commitment and organise pursuit of a clear and compelling vision. The 'Executive Leadership' builds enduring greatness through a paradoxical combination of personal humility and professional will. Collins leaders are "ambitious, but their ambition is first and foremost for the cause, for the organization and its purpose, not themselves." 'Reputational capital' which a celebrated leader brings often give intangible assets to an organisation.

The future world will be much different than the present world. The future of work will also be effected accordingly. One often asks, will there be enough jobs in the future world. This fear is due to the large scale introduction of automation, robotics, and artificial intelligence at our workplaces. Nature of jobs will be altered. There will be more of self-service digital platforms.

The estimates of World Economy Forum (WEF) on 'How technology will change the future of work' indicates that "some 65% of children entering primary schools today will likely work in roles that don't currently exist." Most effected will be office and administrative functions. Business and financial operations along with computer and mathematical functions will see steep rise. The central driver for many of these transformations obviously will be technology.

New technologies enabling remote working, co-working space and teleconferencing will be the principal drivers of change. Our future space of work, says WEF, will be "interconnected workspaces not tied to one place, but many. They will be underpinned by virtual conferencing, complete and constant connection and portability."

Home-working, enabled by technology, will be better defined. "The ICT underpinning these technologies, in consort with the transformational power of big data, could support smart systems that will help tackle climate challenges. Connected homes, factories and farms leveraging smart energy management systems could mean dramatically lower energy use, which would contribute to the decarbonisation of our economies," is the way WEF looks at the future of work vis-à-vis technology.

James Manyika (12) sees major shifts in the job market. He presented "the multiple trends and forces buffeting the world of work drawing on recent research by the McKinsey Global Institute and others." The trends are:

1. Job market challenges are growing, household income stagnating, skill gaps increasing.
2. High unemployment and underemployment, crisis of the kind of untapped human potential.
3. Globalisation has benefitted emerging economies, but has impacted a few jobs moving offshore.

4. Cross-border migration boosts global productivity, but its consequences are often feared by native workers.
5. Scale and speed of technology automation has the potential to disrupt the world of work.
6. Digitally-enabled independent work is on the rise. The reasons cited are want of traditional jobs, supplement income, flexibility needs, etc.

Technology and the Future of Work: The State of the Debate, a report prepared by the Roosevelt Institute for the Open Society Foundations (13) asks - Is Technology a Driver of Wage Stagnation? Is there a race between education and technology, asks Claudia Goldin and Larry Katz (14) .

There was a time when "the best way to ensure a relatively equal society was to educate more people to fill a rising demand for skilled labour." It was the time when education needed to stay ahead of the technology curve. The job polarization theory suggests that with education, creative workers will continue to have strong job prospects; demand for their work is strong, and wages should remain high. But for workers without a proper education, the prospects are dim. While concerned about the downward impact on wages and the rise in low-income work, job polarization scholars remain optimistic about technology's impact on unemployment. They believe the economy will continue to create good, high-wage jobs, and education could provide a pathway to stable employment for many.

Indian Workforce

India has the largest young workforce, it is a growing economy today, yet it is not creating enough jobs. To overcome this situation the country needs to recognise certain fundamental realities, write Samir Saran, Vivan Sharan (15).

The young Indian workforce is expected to carry the mantle of the Fourth Industrial Revolution. According to one survey, "While over 41 percent of the population between 18 and 24 years are already part of the workforce, the others will be joining the workforce in the next two decades." Few observations are:

(1) India has young workforce, under skilled and largely informal. There is low female participation.

(2) Services sector is much less labour-intensive. Though the IT and ITES sub-sectors are the backbone of the services sector, they are understandably not as labour-intensive as factory floors.

(3) There is need to enliven the manufacturing sector, balancing both the low-end manufacturing and also the Fourth Industrial Revolution.

(4) Services sector is not as capital intensive as manufacturing sector is. The probability of job creation is more in services sector.

(5) Keeping in view of the future technological changes the workforce requirement will be of much higher quality than is required in the services sector.

(6) In India informal economy (characterised by the prevalence of micro businesses and low value addition, low levels of productivity, low wages) accounts for a large share of employment creation, in both the organised and unorganised sectors of the economy.

(7) The private sector has to take greater responsibility in the creation of productive workforce. The primary role of government will be to cater the needs of the sectors "where the average revenue per user is among the lowest in the world, but the user base is among the largest in the world."

The OECD initiative on the future of work looks at how demographic change, globalisation and technological progress are affecting job quantity and quality. OECD's digital report reflect a few trends that would be helpful in the design of future workforce.

(1) 14% jobs could be automated with 32% likely to change significantly. This is unlikely to reduce job opportunities. Technology will make some occupations obsolete; it will also create new jobs.

(2) 6 out of 10 adults lack basic ICT skills. Share of highly-skilled jobs has increased by 25% over last 2 decades, low-skilled jobs have also increased, but the share of middle-skilled jobs has decreased.

(3) Between 1995 and 2015 employment in the manufacturing sector went down by 20%, it rose by 27% in the service sector.

(4) Self-employment, part-time and temporary work are on the rise. They represent around 14% of total workers across OECD countries. They are usually less well covered by social protection.

(5) Technology can improve work-life balance, can create new opportunities, can automate tedious and dangerous tasks, can improve health and safety, and can boost productivity. This will require new learning skills.

ILO proposes a human-centred agenda for the future of work, based on increasing investment in people's capabilities, institutions of work, and decent and sustainable work.

SMART machines are important for the future of work. Shoshana Zuboff 's book (16), In the Age of the Smart Machine: The Future of Work and Power, written nearly three decades ago suggested that "to fully grasp the way in which a major new technology can change the world it is necessary to consider both the manner in which it creates intrinsically new qualities of experience and the way in which new possibilities are engaged by the often conflicting demands of social, political, and economic interests in order to product a choice." The stand on this subject has not changed much.

The ten levers of smart engineering, as enumerated by my INAE colleague are: 'Need of the Hour Engineering' (based on the premise that each era's emphasis on the aspect of engineering is different), 'Improvised Engineering' (how same or similar purpose is achieved by more sophisticated technology), 'Strip Down Engineering' (a combination of reverse engineering and frugal engineering to give an improved product), 'Performance Boosting Engineering' (enhancing possibilities keeping constraints in mind), 'IntelliSys Engineering' (intelligent systems to improve autonomous operations), 'Cross Pollination Engineering' (recognition of the fact that solutions for some problems require extensive knowledge of multiple faculties), 'Smart Auxiliary Engineering' (support role of engineering should not be misconstrued as engineering trivia), 'Sustainable Engineering' (play, if possible, zero sum game with the environment), 'Nature Inspired Engineering' (take lessons from the nature to design a product), and 'Forward Looking Engineering' (keeping in view the fusion of the physical and the digital world).

Evgeny Morozov (17) seems worried about the 'problem solving power of our technologies'. With the alarming increase of this power, Morozov thinks, there exists some possibility of diminishing our ability to distinguish between important and trivial or even non-existent problems. There is no need to fix every conceivable problem just because we have smart solutions. There is no need to take advantage of smart technology's every conceivable possibility.

'Smart' makes us more plastic, more programmable. Morozov's worry is that "blinded by the awesomeness of our tools, we might forget that some problems and imperfections are just the normal costs of accepting the social contract of living with other human beings, treating them with dignity, and ensuring that, in our recent pursuit of a perfect society, we do not shut off the door to change." Sterile and contended environments are not well-known for innovation.

This divides us into two camps. Paul Saffo (18), Technology Forecaster at Stanford University, calls them 'Druids' and 'Engineers'; one thinks we must slow down while the other thinks technological innovations can solve all problems; one wants to return to the past while the other wants to fly to the future; one wants to ban the GMO while the other wants to create life; one thinks robot cars are unsafe while the other wonders why humans are allowed to drive at all; one is an optimist (anything can be fixed if there are enough brain, effort and money resources) while the other is pessimist ("no matter how grand the construct, everything eventually rusts, decays and erodes to dust"). Saffo's advice: we must resist the pressure from either side. We can't revive the past neither can build technologies that don't carry hidden trouble.

The Economist recently asked an important question — whether the creation will be worth the destruction. This question is raised, keeping in mind, the impact of smart machines. According to a study made by the McKinsey Global Institute, the dramatic progress that we are seeing is due to a combination of Moore's law and the melding of three technologies: machine learning, voice recognition and nanotechnology. The study argues that extraordinary developments will make knowledge workers more productive, and will be helpful for both entrepreneurs and consumers. The study points out justified worry and that is that modern technologies will widen inequality, increase social exclusion and provoke a backlash. The Economist writes: "Innovation will disrupt many areas of skilled work that have so far had it easy. But if we manage them well, smart machines will free us, not enslave us."

Society 5.0

Yuval Noah Harari (19) describes human history (considering human species as a single data-processing system, with individual human serving as its chips) through four basic methods. (1) As population increases, the number of processors increases, thereby improving the computing power. (2) There is corresponding increase in the variety of processors, hence its dynamism and creativity and generation of novel ideas. (3) Possibility of increased number of connections between processors that are likely to result in better economic, technological and social innovations. (4) Increased freedom of movement along existing connections; just building roads is not useful if plagued by robbers. These data-processing systems passed through four main stages. (1) The first stage of cognitive revolution gave crucial advantage to human species in comparison to non-human species. This resulted in the increase in the number and variety of human processors. (2) The second stage of agricultural revolution accelerated demographic growth and networked living. Centrifugal forces remained dominant; humankind still divided. (3) The third stage of writing and money overpowered the centrifugal forces by the gravitational pull of human cooperation. Human groups bonded and merged to form cities and kingdoms. (4) The fourth stage of scientific revolution encompassed the whole world. There was flow of information throughout the global grid. There was considerable improvement in data-processing system. In the span of 70000 years we have crossed various ages. We have moved from hunter-gatherers society (Society 1.0) to agrarian society (Society 2.0) to industrial society (Society 3.0) to information society (Society 4.0). We are now entering Society 5.0.

Society 5.0 is a super-smart society. Society 5.0 aims at creating a society of the future incorporating the innovations of the fourth industrial revolution. In this society, Big Data collected by IoT will be converted into a new type of intelligence by AI. It aims to make people's lives more comfortable and sustainable. Japan is the leader of this movement because it is data rich country. Japan will play a key role in expanding the new Society 5.0 model to the world.

Society 5.0 promises to change the world. Take the issues of increasing medical and social security expenses of the aging society of Japan. Society 5.0 envisages following solutions: By connecting and sharing dispersed medical data, effective medical treatment based on data would be possible. Remote medical care will makes it possible for the elderly people to initiate treatment sitting at home. Many areas lack access to public transportation due to the shortage of drivers. This problem can be sorted out by promoting autonomous driving taxis and buses for public transportation. Delivery drones can improve distribution and logistics efficiency. Sensors, AI and robots will be used to inspect and maintain roads, bridges, tunnels and dams. Society 5.0 will use blockchain technology for money transfer, and shall promote cashless payment.

All this will obviously impact the future of work.

What do we do with our time?

A prediction made in 1930 by a prominent economist was that by 2030 we will be entering 'the age of leisure'. The ordinary people will have so much time that it will be difficult for them to spend it. This prediction has not come true. We have become so 'busy' that we have no time to think. This raises a pertinent question – what do we do with our time? We call ourselves 'busy', are we really busy? How over the years the pattern of time have changed? These and many such questions are deliberated by Jonathan Gershuny and Oriel Sullivan (20). One of their observation is that the regular pattern of eating breakfast, lunch and dinner has almost vanished. People are developing irregular eating habits; they are eating and snacking throughout the day. Another of their observation is that the time we spend on leisure and sleep has changed relatively little. Moreover, we tend to overestimate the amount of time we spend on work. Yet another observation is that the amount of time women spend in paid work has gone up and the amount of time they spend doing domestic work (cooking, cleaning and laundry) has dropped substantially. On the other hand, men are spending more time doing domestic work, though far less than women. It is also a myth that people are much busier than in the past. Highly educated in higher status jobs are perhaps busier than before. Their voice carries weight, and therefore, creates a general impression that everyone is now busier. 'I am busy' enhances status and signals indispensability.

We have asked many questions. Let us now stop. Let our neural suitcase decides the size of our world. We live in a 'pale blue dot', the only dot that offers life.

At time zero,

When the dot was not pregnant with life,

Seed was needed for life to begin at the dot.

The first cell, no one knows, from where it arrived,

Started to divide and multiply.

The cell wanted to become better,

Thus began the journey of evolution.

It brought chaos and disorder.

To contain disorder, and to maintain order,

Nature wrote the script;

The script of a pale blue dot.

The only dot that offers life!

Let our Science-Engineering-Technology mindset keep the dot alive!

7

COVID Crisis and Technology Update

The manuscript was completed in February 2020. The book was supposed to be out in print form in April 2020. In between the pandemic changed the world. Obviously, world changing events can't be left out of the book, particularly if the book is about the future of science, engineering, technology and humanity. I, therefore, requested the publisher of the book for the inclusion of a chapter COVID crisis and its impact on the future of the world. He readily agreed.

I am sincerely grateful to the publisher for allowing me to include issues relevant for the present book, as an additional chapter at the end of the book. The subject at the moment looks unending and it has so many dimensions. I have tried to cover them as briefly as possible, in the context of the book. In spite of the several difficulties due to this last minute inclusion the publisher has taken pains to include it. I am grateful to him for his gesture.

The pandemic of this proportion is expected to get the views/suggestions from equally large number of people. Everyone is eager to see the end of the pandemic. And everyone has realised that the world will be a much different place. Some people are hopeful that the problems will be resolved, of course using multi-dimensional approaches. Not many are sure of the time frame. There is also pessimism among common people. One of my aunt's asks me - what kind of research we have been doing all these years that we can't even face the challenges posed by a mere virus? Has the balloon of the technology busted by a miniscule virus? In this fluid time many questions are being asked, and as many answers are also emerging. It is a big issue. It is as big as that emerged after the Second World War. Some are calling it the 'birth of a new power'.

We need to understand a few basic issues, like - What are its likely social, economic and technology impacts? How good is our preparedness to deal with the pandemic? Once the storm is over are we ready to face this 'new normal'.

Before we understand the situation, few things come to my mind:

1. The world will need a different kind of preparedness to deal with unseen problems.
2. Minimalism and optimum optimism is the need of the day.
3. Black swan, premortem, and scale up are essential matters for any discussion onbig projects.
4. World bodies will have to adhere to universal ways more rigidly to deal with pandemic like situations.
5. Slums of the world should get a facelift, not only to reduce the spread of infection, but also to reduce the chances of getting infected, not only from the viruses but also from other vices.

The question is - Is Coronavirus the 'Black Swan' of 2020? Before I respond to this question, take a look at the question-: Is coronavirus man made?

Coronaviruses are known to cause illnesses ranging widely in severity. The first known severe illness caused by a coronavirus emerged with the 2003 Severe Acute Respiratory Syndrome (SARS) epidemic. The second outbreak of severe illness, known as Middle East Respiratory Syndrome (MERS), happened in 2012. In December 2019, the World Health Organization (WHO) alerted the world of an outbreak of a novel strain of coronavirus causing severe illness, which was subsequently named SARS-CoV-2.

The studies published by Kristian Andersen and his colleagues in the journal Nature Medicine (1) say that SARS-CoV-2 is the result of the natural process of evolution rather than a product of biological engineering. The observation is based on the analysis of available genomic data. Initially it was thought that the virus originated from the seafood market in Wuhan. It was also suggested that the virus may have spread to humans from illegally trafficked mammals pangolins.

Andersen and his team looked at the spike protein that is used to bind coronavirus to the membrane of the human or animal cells that they infect. Two components of spike proteins were studied: the receptor-binding domain (RBD), which latches onto healthy host cells, and the cleavage site, which opens up the virus and allows it to penetrate the host cell. To bind to human cells, spike proteins need a receptor on human cells called angiotensin-converting enzyme 2 (ACE2). The scientists found that the receptor-binding domain of the spike protein had evolved to target ACE2 so effectively that it could only have been the result of natural selection and not of genetic engineering. The backbone of SARS-CoV-2 is very different than those of other coronaviruses and is similar to related viruses in bats and pangolins.

Typically, if a researcher was engineering a new coronavirus as a pathogen or biological warfare agent, they would build it on the backbone of a virus that is already identified as causing disease. In the case of SARS-CoV-2, research showed that the SARS-CoV-2 backbone was quite different from known coronaviruses and more closely resembled viruses found in bats and pangolins. They thus ruled out the possibility of laboratory manipulation. One of the scenarios presented by the claimants is that the virus evolved to become pathogenic in an animal and then jumped to humans. They agree to the theory that bats are the most likely carrier, as SARS-CoV-2 is very similar to a bat coronavirus. It could also be a possibility that the virus was transmitted to humans from bats. It could also be possible that the virus is non-pathogenic in animals but jumped into humans and evolved into a disease-causing strain there. The researchers, however, don't vouch for any of the hypotheses.

Shortly after the epidemic began, the genome of SARS-CoV-2 was sequenced. The data was made available to the researchers worldwide. Andersen and collaborators at several other research institutions used this sequencing data to explore the origins and evolution of SARS-CoV-2 by focusing in on several tell-tale features of the virus. (2)

Coronavirus is not the black swan of 2020, it is a gray rhino, writes MicheleWucker (3). She thinks, "And the facile willingness to see crises as black swans has provided policymakers cover for failing to act in the face of clear and present dangers from climate change to health care to economic insecurity." Wucker prefers to call it "gray rhino"; The gray rhino is the massive two-ton thing with its horn pointed at you, stomping the ground and getting ready to charge — and, most important, giving you the chance to act.Some say, we should have heeded to past warnings; if we had, we would not have to face the consequences we are facing today. There were ample clear warnings. It is said that we have similar threat signals, for example, from climate change and antibiotic resistance. 'Gray rhino' approach is most helpful to deal with this kind of problems, thinks Wucker, and then she adds, "The black swan has one benefit, which is that it encourages people to expand their ideas of what might happen."

The coronavirus is a bad fit for NassimTaleb's definition of black swan, thinks James Pethokoukis (4) "But is the emergence of such a dangerous virus really an unpredictable outlier that suddenly swooped in from outside our 'regular expectations'?" asks Pethokoukis. What NassimTaleb says "... some people claim that the pandemic is a "Black Swan", hence something unexpected so not planning for it is excusable. Had they read that book (Black Swan), they would have known that such a global pandemic is explicitly presented there as a white swan: something that would eventually take place with great certainty.

Such acute pandemic is unavoidable; the result of the structure of the modern world; and its economic consequences would be compounded because of the increased connectivity and over optimization."

COVID Crisis: Biology and Technology

David Cyranosk (5) while presenting the 'profile of a killer' and its complex biology writes, "The latest corona virus, named SARS-CoV-2, has evolved an array of adaptations that make it much more lethal than the other coronaviruses humanity has met so far." Genetic evidence suggests that this virus has been hiding out in nature possibly for decades. "I've been working on coronaviruses for 20 years, and most of the time it was neglected and not taken seriously. Now that it's happened, how can I leave?" says Structural biologist Rolf Hilgenfeld. He was to retire on 1 April. He is still chasing the structure of the SARS-CoV-2 virus.

Corona viruses are big in size (125 nanometres in diameter) and have the largest genomes of all RNA viruses, having 30,000 genetic bases. The latest corona virus genomes are more than three times as big as those of HIV and hepatitis C, and more than twice influenza's. "A neighbour's cough that sends ten viral particles your way might be enough to start an infection in your throat, but the hair-like cilia found there are likely to do their job and clear the invaders. If the neighbour is closer and coughs 100 particles towards you, the virus might be able get all the way down to the lungs." (5)

The complete genome of a coronavirus causing a cluster of pneumonia-like cases in Wuhan, China, was posted online on 10 January, 2020. "Within 24 hours, a network of structural biologists around the world had redirected their labs towards a single goal — solving the protein structures of a deadly, rapidly spreading new contagion. To do so, they would need to sift through the 29,811 RNA bases in the virus's genome, seeking out the instructions for each of its estimated 25–29 proteins. With those instructions in hand, the scientists could recreate the proteins in the lab, visualize them and then, hopefully, identify drug compounds to block them or develop vaccines to incite the immune system against them." (6)

On 11 January 41 confirmed cases of COVID-19 worldwide were reported, on 13 January 42 confirmed cases, on 26 January: 2,014 confirmed cases, 1 February: 11,953 confirmed cases, 18 February: 73,332 confirmed cases, 16 March: 167,515 confirmed cases, 22 April: 2,471,136 confirmed cases, 14 May: 4,248,389 confirmed cases.

The human coronaviruses mainly cause respiratory infections. Different people have different experiences with coronavirus. It can produce cough disrupting taste and smell. It might work its way down to the lungs and debilitate that organ. Most infected people create neutralizing antibodies that are tailored by the immune system to bind with the virus and block it from entering a cell. But some people seem unable to make them.

Testing Covid cases and developing vaccine for treating COVID cases are the two important ways to curb the Covid menace.

The Journal Nature (7) examined Coronavirus tests that are currently available and the status of new diagnostic test that are being developed. Most tests are based on reverse transcription polymerase chain reaction (RT-PCR) using material from nose and throat swabs. The method amplifies a specific gene sequence in the virus. Primers are used for copying purposes. Different labs use different primers for targeting different sections of the virus's genetic sequence. PCR-based tests have often given false positives, due to contaminated reagents used in the tests. Key requirements for conducting the test are appropriate lab space, reagents and tools and trained people. Often trained people are in short supply. Most countries were not prepared to take up the test, due to various reasons. Therefore, barring a few countries, the pace was slow in adopting and conducting the tests. Scaling of test facilities, obviously, is one of the reasons of slow development of testing facilities. Another option is serological test. This test looks for antibodies in previously infected people to detect infection. Several groups are working on this option, no test as yet been broadly validated for clinical use. Some groups intend to use CRISPR that take advantage of the popular gene-editing technique to improve testing. "The techniques use the CRISPR machinery's ability to recognize specific genetic sequences and cut them. In the process, it also cuts a 'reporter' molecule added to the reaction, which reveals the presence of viral genetic material relatively quickly," writes Nidhi Subbarama (7). She adds, "The key advantage is that a CRISPR reaction is incredibly specific and can be done in 5–10 minutes."

How the Coronavirus Acts?

Coronavirus uses its surface spike protein to lock on to ACE2 receptors on the surface of human cells. Once inside, these cells translate the virus's RNA to produce more viruses (8). Callaway has explained the process in the following 5 steps: 1. Virus enters the body. 2. Virus enters the cell. 3. Virus fuses with vesicle and its RNA is released. 4. Virus assembly. 5. Virus release.

How the Immune System Works?

The antigen-presenting specialized cells engulf the virus and display portion of it to activate T-helper cells. T-helper cells enable other immune responses: B cells make antibodies that can block the virus from infecting cells, as well as mark the virus for destruction. Cytotoxic T-cells identify and destroy virus-infected cells. Long-lived 'memory' B and T cells that recognize the virus can petrol the body for months or years to provide immunity.

How the Vaccine Works?

Vaccines provoke an immune response that can block or kill the virus in infected person. Based on the following strategies more than 90 vaccines are at different stages of development.

1. Virus vaccines – in weakened or inactivated form, as used in many existing vaccines. One company in the US is working with a company in India using weakened viruses by altering the viruses genetic code so that less effective viral proteins are produced. Virus is made ineffective using chemicals or heat in the inactivated form.

2. Nucleic acid vaccines – DNA and RNA are used for genetic instructions that prompt a immune response. Inserted nucleic acids churn out copies of the virus protein that encode the virus spike protein. Their practical utility is yet to be proven.

3. Viral-Vector vaccines – a known disease producing virus is genetically engineered to produce coronavirus proteins in the body. These engineered viruses are weakened. Two types are available: those that can still replicate within cells and those that can't. Replicating viral vectors are safer and provoke strong immune response. However, previously existing immunity could blunt the vaccine's effectiveness. Non-replicable viral vectors need boosters for long-lasting immunity. Viability of this method yet to be proven.

4. Protein-based vaccines – injecting coronavirus proteins directly into the body; focus is on spike protein or receptor binding domain. Virus like particle mimic the virus structure but are not infectious. They can, however, trigger a strong immune response. They seem difficult to manufacture.

Vaccine design is the priority of many R&D institutions, universities and companies. India is not lagging behind. According a report, substantiated by the Department of Biotechnology and Department of Science and Technology, Government of India, several big government-backed projects — in both private firms and academic institutions — are leading the country's Covid 19 vaccine hunt. (9). One of the projects has entered the trial stage.The research firms are: Serum Institute of India, Pune; Cadila Healthcare Ahmedabad; Bharat Biotech, Hyderabad; Gennova, Pune; CMC Vellore; National Institute of Immunology, New Delhi; IIT Indore; Enzene Biosciences, Pune; IICT Hyderabad; Aurobindo Pharma, Hyderabad; Seagull Biosciences, Pune; IISER, Mohali and Trivandrum. Secretary DBT, GOI said: "Over the years, India has emerged as the global vaccines hub. Several top vaccine giants in India supply basic and advanced vaccines to about 150 countries. India has earned the recognition of having the largest global capacity for WHO prequalified global manufacturing."

In addition to vaccines, dozens of coronavirus drugs are in various phases of development (10) Remdesivir (previously used for Ebola virus) is the most talked-about drug. Trials indicate reduction in recovery time by a few days. Familiar generic medications, such as hydroxychloroquine, are repurposed for coronavirus treatment. Antibody treatments have also been used to "tamp down the body's immune response when it becomes destructive"

The scale up is the most likely engineering challenge of each drug manufacture in large scale. You need many things to make a thing. Production chokes due to the non-availability of even the smallest thing. That is the challenge of scale up of supply chains. Scale up is a problem for all molecules, big or small. Scale up problem we faced for the mass production of penicillin, not long ago, and we solved the problem of scale up. . We will solve the current problem too.

When you have so many vaccines in race, it is difficult to choose the winner. (11) Developers and funders are laying the groundwork for efficacy trials to only a handful of vaccines. Then there is 'human challenge'. Large trials are necessary to determine safety and efficacy.

India took Covid-19 'very seriously' from the start, says WHO chief scientist Soumya Swaminathan. She also called for an 'open mind' about traditional medicine. She said "ICMR trusts science," and added that their data can be believed. "I do not believe they would fudge data or hide information." She is of the opinion that 'Asia contained Covid-19 better than European nations'. Soumya Swaminathan expressed support for using traditional medicine to boost immunity. Traditional medicine, she said, should be given a fair chance in either preventing Covid-19 infections or for treatment.

India is currently testing 75,000-80,000 samples a day through the reverse transcriptase polymerase chain reaction (RT-PCR) test, the gold standard for confirming Covid-19 cases , and has set itself a target of testing at least 100000 samples per day by the end of May. (12). Pune-based Mylab Discovery Solutions has manufacturing capacity sufficient to meet current needs of the country. They have a tie up with Serum Institute of India Ltd.

Indian Council of Medical Research (ICMR) and Pune-based National Institute of Virology (NIV) have developed Elisa based antibody testing kit for coronavirus. The first batch of completely indigenous IgG ELISA test kit is ready and has also got ICMR clearance, said an official statement. ICMR has signed a 'non-exclusive agreement' with Zydus Cadila to take up the production of ELISA. The NIV, Pune has validated the first batch of ELISA kits produced by ZydusCadila.

COVID Crisis: Education

Though we love to project gloomy picture, any crisis has two sides. We often forget about the rosy side. We forget wherever there are challenges there are also opportunities. Future of education in the context of current crisis seems to lie on both the sides, depending upon who is steering it.

We have always faced learning challenges. We are never happy with what is given to us on the platter. When we get a good view of, say for example, physics, chemistry and mathematics we are told we arenot learning the fundamental skills needed for life. This pandemic has potential of worsening the situation, unless we are extra alert. We must think but also we must act before it is too late. The good thing is that lot can be done to reduce the impact of the crisis, but that will require, besides infrastructure, lot of efforts of the society, the government, and of course the major stake holders. All the strata of the society will be affected, middle-income and poor-income groups more. If we are careless we will allow inequalities to amplify. What we must do is to minimize the differences in opportunities, as much as possible.

"The Covid-19 pandemic offers universities a once-in-a-generation opportunity to put their dysfunctional strategies behind them," argue Timothy Devinney and Grahame Dowling (13). The other storytellers say that the crisis is a major threat and there will be "catastrophic shortfalls in university revenue, which will lead to massive job cuts and severe disruptions to learning and research." Both sides are saying the truth, depending upon how the truth is interpreted. Devinney and Dowling see "most public universities look more like bloated conglomerates than focused intellectual-capital and information-dissemination institutions that can help the economy and society navigate the future."

Running an education system costs money. On the other hand, expectations from it are much more than 'making money' and 'students getting jobs'. There is nothing wrong in filling the beds provided one is careful about the treatment. We are facing a 'concerned' generation. And definitely there is nothing wrong in students getting lucrative jobs.

Higher education has always been a matter of concern for many countries. There have always been the issues of content, investment and output. There are problems of expectations from this enterprise, particularly if it is privately funded. The understanding of cost control is a matter of opinion. Variation in understanding between the cost controller and the implementer can be so wide that the operationalization of the enterprise may lose its meaning.

Generation of revenue by reduction of academic support is possible, but not desirable and acceptable. Accountants can't run an academic enterprise. Imparting education and maintaining account book are two different ball games. Piecemeal efforts make sense only in the short run. 'For lambi race kaghoda' short term fixes often become long-term addictions, and they do what addictions do. They lead to long-term structural problems. Mushrooming of low quality engineering and business schools prevailing all over the country can lead to long-term structural problems.

At this point I must say that starting a business school and an engineering school are two entirely different situations; one doesn't need capital resources, while the other needs lots of it, besides human resources. The requirements to teach at these schools are also different. Too much focus on specialization at the undergraduate level needs lowering. Matters of red tape can be resolved amicably only if that smoothens the process of working. It is for the management to see if they want 'red tape' or not.

Another question is who should govern – an academic, an accountant, or a manager? If 'one is three' that is the best solution. But that seldom happens; their ways of governance, due to the nature of their training, are so different. I have often seen that an academic at the peak of excellence is offered the position of the head of the institution. The problem is that we all want 'corner room'. We deserve it in some senses, while in many other senses we are so undeserving even we don't know.

Can't we have more corner rooms? Bureaucracy stifles creativity. Accountant sees only one colour. Administrator's job is to run the shop. He has to fight with the authorities with one hand. He has to keep the other hand ready to fight with the government. Leadership roles need to be clearly redefined. One must not waste time on an activity she/he is not fit to handle.

I am not saying that education should be free. What I am saying is that students should be willing to pay. Students will willingly pay only if the education provided is of the first quality. There should be places to go to, if one wants financial support. Students have the right to know what they are going to get. One expects to get what one pays for. More than the number of students, it is the quality of the students that is bothering some educators.

There is a need to understand that 'premortem' is as essential as 'postmortem' is. We need to train 'locals'. Like in many other countries, we can't depend upon the foreigners for revenue resources. We need many peaks, instead of one or two at the top. As they say, buying education and buying meat are two different things.

The meaning of 'learning', per se, has changed. Materials for learning are available at our tips. We need to train students who don't know what kind of job they are going to get, because of the uncertain future of the 'future'. According to one report, 85% of the jobs in 2030 have not yet been invented.

One may agree that education is for all. One may also agree that running an education enterprise is not everybody's cup of tea. Not every professor is a director material. An accountant must make the habit of seeing 'green', besides the usual 'red'. An education administrator must understand that she/ he is dealing with humans, not mere mechanized products of an enterprise. An administrator must understand the importance of colour brown; a mix of red and green.

On-line education is one solution for the delivery of education, but for that we need huge resources. Our aim is not to teach only those who can afford education. Elite schools can afford all kinds of educational modes of imparting education, crisis or no crisis. They have different kind of incentives. What we need is the provision and placement of an appropriate support system. Education is also for the poor. Will social media and mass communication systems come forward to promote not-for-profit activities? We have to think if our drop-out rates are high. All our efforts, of course, will depend upon what we are going to do with education.

We are observing a change in our attitude. It is for the better. An attitude of help is coming back. It is heartening to know that companies are temporarily shifting their products base from their regular supplies to something with the sole objective to overcome the paranoia of the pandemic. During a crisis, it is good to note, a massive spike in energy present in the workforce.

COVID Crisis: Technology Innovation

Vision of engineering in the new age envisages myriad challenges. Some of the best practices in engineering education, according to a MIT report are:

1. The challenge of delivering high-quality, learner-centred education to large and diverse student cohorts;
2. The 'silo' nature of many engineering schools and universities that inhibits collaboration and cross-disciplinary learning;

3. Faculty appointment, promotion does not appropriately prioritize and reward teaching excellence;
4. The quality of education can't be measured merely on the basis of staff-to-student ratios, and graduate employment profiles;
5. "The scholarly work going on in engineering education is not translated back into the lecture room, it's always theoretical";
6. The discipline/department-based structure of many engineering schools and universities are holding back innovation and excellence in engineering education;
7. The alignment between governments and universities in their priorities and vision for engineering education, like purpose of engineering undergraduate education, government regulation and national accreditation, unpredictable nature of higher education funding, and commoditisation.

Several questions related to higher education bother us. What type of skills should institutions foster through higher education? How should skills be taught in order to ensure the best education outcomes? How should universities fund themselves and their research, and how important is their research to modern-day higher education? Should institutions tailor courses to the needs of employers or to the requirements of students? Should universities emphasise civic engagement as well as educational performance in order to address societal divisions?

No one expects simple answers to these questions and also universal answers. Expenditures to run academic institutions are rising, and that is changing the perspective of higher education. A report by The Economist Intelligence Unit discusses these 'existential questions'. There is general agreement that higher education needs to evolve, but the question is how, and particularly in the post COVID world. The report highlights five innovative models of higher education.

1. Offer higher education "anytime, anywhere, to anyone" through Online Universities. Providing online learning to marginalised students is not easy.
2. The cluster model fuses multiple institutions thereby sharing services and facilities and reducing overall costs and offering students a greater range of course options. The success of such a venture will depend upon the bonhomie among the cluster institutions.
3. Experiential institutions bring teaching out of the classroom. Learning is driven by diverse experiences such as internships or hands-on projects.

Though experiential in nature, this model has little operational experience. Asking a conventional teacher to move away from traditional lecturing is not easy.

4. Liberal arts college models are somewhat like personalised education models where the operational costs are high due to lower student-teacher ratio. While some argue that wide-ranging generalist skills are increasingly needed in the corporate world, but others hold that institutions must focus, at least to some degree, on more targeted pre-professional skills.
5. The partnership model works on joint venture in order to secure long-term funding and improve job prospects for graduates. Courses with focus on upskilling and reskilling and are tailored to particular jobs or skill gaps that partners face. Obviously this model limits the scope of education. Moreover, it makes the students prisoners of industry partner. The purpose of industry is not to do charity.

On research W. I. B. Beveridgewrote years ago: "The most important instrument in research must always be the mind of man." Are we using the 'instrument' differently than our predecessors?

While we know about known unknowns, there are also unknown unknowns. Cognitive scientist Gary Marcus says, unclear risks that are in the distant future are the ones we take less seriously. We generally discount the future impact of the risks, because we don't know them. We like to believe that we live in a world that has fewer unclear risks. Often, unclear risks cause more serious problems than clear risks. We often ignore the future risks. We are forced to remember them only after the 'risks' happen. Says Marcus, "What we really should be worried about is that we are not quite doing enough to prepare for the unknown."

Alex Osborn observed more than six decades ago, "It is easier to tone down a wild idea than to think up a new one." How effective are Osborn's concept in the present environment? Take a look! When people want to extract good ideas from a group, they still obey Osborn's cardinal rule, censoring criticism and encouraging the most freewheeling associations. Several recent studies, however, have exemplified 'sobering refutation of Osborn'. Studies have indicated that brainstorming makes individuals less creative. They say fewer ideas are generated when people pool their ideas, compared with when they work alone. But then there are many who believe creativity is a group activity.

R Keith Sawyer unfolds some hidden secrets of creative minds. He says, there is nothing like a "full-blown moment of inspiration." Ideas don't magically appear in a genius' head from nowhere; they always build on what came before.

An idea may seem sudden, but in reality our minds have actually been working on it for a long time. Ideas emerge from a chain reaction of many tiny sparks. Often we fail to gauge which spark is the brightest. Advice of Sawyer is to develop a network of colleagues and schedule time for freewheeling and unstructured discussions. Look at what others in your field are doing. Brainstorm with people in different fields. Distant analogies often lead to new ideas.

When you focus your attention on something, you only see a very small fraction of your field of vision, because your brain gets filled in with everything what you think is there. In doing so you may miss the disruptive things that are happening at the periphery. Says Joichi Ito, "You've got to be antidisciplinary, because if you're in a discipline and you're worried about peer review and you're knowing more and more about less and less, that's by definition an incremental thing." Ito says, when you are anti-disciplinary you have the freedom to connect things together that aren't traditionally connected. Some of Ito's prescriptions are: stop focusing on individuals and start focusing on communities; stop focusing on top-down and focus more on bottom-up; stop focusing on single experts and start focusing on the cloud. The other point Ito conveys is that you may not have a first class degree, but you know what really you are good at, and what you are obsessed with. What Ito conveys is that to steer the innovation boat you also need few 'misfits'.

Is creativity the preserve of a few geniuses or can be found, in some form, in all individuals? Can we just give somebody a bunch of free time and then expect him to come up with a brilliant idea?

An emerging element of innovation is openness. Open innovations go out to find solutions. It doesn't matter if solutions are available with a competitor; an open innovator makes all efforts to obtain it, not necessarily by adopting buying strategies.

Rex Jung studied the interaction between creativity and intelligence, mainly among college students. Jung says, "Creativity and intelligence are linked at lower levels of IQ, but above a certain threshold, they don't necessarily go hand in hand." Among the ways suggested to foster creativity, besides purpose and intention, are building motivation, developing self-management and leading a fulfilling life.

Howard Gardner thinks that talent and expertise are necessary, but not sufficient to make someone original and creative; "achievement is not just hard work: the differences between performance at time 1 and successive performances at times 2, 3, and 4 are vast, not simply the result of additional sweat."

We are often reminded about the low percentage of engineering graduates that are employable. The reasons put forward are profit-hungry management, lack of skill education, resplendent corruption, focus on rote-learning, and shortage of faculty (both in quantity and quality).

What we need is to be innovative and socially responsible leaders. Top employers search for exceptional scientists who can bring fresh and original ideas to the company. It is not enough for an innovative company to hire people who are only exceptional scientists. The person should also fit into the company's core values. An innovative company recognises the joys of uncertainty. Maintaining status quo doesn't satisfy an innovative company. In order to attract the right people, for example, one company advises the job seekers not to apply, if science is not their obsession, if one is content being the smartest person in the room, and if one is afraid to fail.

In other words, if you are routine, you are not fit to work in an innovative company.In a dynamic environment of the future, the future engineers are expected to learn continuously throughout his or her career, not just about engineering, but also about history, politics, business, and so forth. It is imperative that the industry joins hand with academic and R&D community in the spirit of togetherness and sense of fulfilment.

India is fortunate to have young driving force who can think 'outside the box'. What are the forces that drive the young to think outside the box? What must we do differently, as parents, teachers, mentors and employers to prepare young minds as innovators?

There is a view that says that teachers make the most critical difference in the lives of young innovators. These teachers are not ordinary teachers, but outliers, identified by the ways in which they teach. The young generation, Wagner believes, can live on less. He, however, adds, "Those are easy things to say when you're in your 20s. When you're in your 30s — thinking about a family — that picture may change". He, however, feels today's generation is less materialistic. This 'connected generation' knows how to find support for what they want and need to do.

In our country we are producing about 9000 science, technology and engineering PhDs every year and hope to produce up to 20,000 PhDs each year. More than the number, the quality of PhDs produced is important. One way of improving the quality is to revise the reward and recognition structure. In one of the Webinars I heard someone saying that we have very large number of part-time PhD scholars compared to full-time. Therefore, large number is a misnomer.

What the students think of doing PhD in India? In one survey, IIT students mentioned several reasons for not doing PhD in India. The reasons included too much time taken to complete PhD, too many pre-PhD courses, low market value, and uninspiring supervisors. Some supervisors are too 'inspiring'; they don't hesitate to supervise more than a dozen students at a time. There is a need to change this trend. In several institutions there is severe resource constraint and poor resource sharing. Let students also understand that completing PhD is their responsibility, not their supervisor's.

What the industry thinks of PhD holders? Many think, PhD holders are poor team members. They are less adept at dealing with changing challenges. Many think some non-academic training is essential for PhD aspirants. They say courses in marketing, communication and leadership are useful for a scientist alongside academic acumen of critical thinking and analysis. An issue that needs consideration is to "trample the boundary" among the scientific disciplines, because of the trans-disciplinary nature of science and technology. The engineering institutions that are established in recent years have carefully excluded disciplines that are no longer relevant for resolving many of the current problems.

Our dealing with the pandemic is failure of our preparedness to deal with the situation, believes Marc Andreessen (14). He thinks it is our "inability to build." He says "We chose not to build." Andreessen dubbed the current crisis the problem of inertia and lack of will. We don't have vaccines and medicines, or even masks and ventilators. It is our lack of foresight and failure of imagination. We are lost in the maze of IT. We are forgetting to keep up with our basic needs amidst 'shiny software'. We are becoming far less accomplished, writes David Rotman (15).

Every manufacturer is impacted by this crisis in some way, and for many this poses an existential threat (16). One is not sure how much time it will take for things to return to some level of normality. Thinking has changed. In the pre-Covid scenario industry 4.0 was focussing on competitive advantage, cost reduction, productivity, sustainability and innovation. The focus now will be on how to run the business better. Survival will be foremost in the mind of manufacturers. Damage control will take precedence. There will be reduction in non-essential spending and investment. Many essential investments will become non-essential.

FORBES (17) predicts the following changes (for the better) in Next-Gen Tech. (1) Digital Product Testing – digitizing the supply chain shortens lead times, aligns products with the right price and quality. (2) Health Robots – to remove all airborne viruses and bacteria on the surface. (3) Work-From-Home - remote work will stay. (4) Ed-Tech – e-learning will supplant normal learning.

Coronavirus will change the world. What some of our experts say (18):

1. The comfort of being in the presence of others might be replaced by a greater comfort with absence, especially with those we don't know intimately.
2. We will finally start to understand patriotism more as cultivating the health and life of your community, rather than blowing up someone else's community.
3. Perhaps we can use our time with our devices to rethink the kinds of community we can create through them.
4. "The reality of fragile supply chains for active pharmaceutical ingredients coupled with public outrage over patent abuses that limit the availability of new treatments has led to an emerging, bipartisan consensus that the public sector must take far more active and direct responsibility for the development and manufacture of medicines."
5. The trauma of the pandemic will force society to accept restraints on mass consumer culture. Our outsized appetite needs restraint. We have to decide how much our engineering footprint should be allowed to grow.

Let's be open to adjustments. Together we can face the unknown unknowns.

COVID Crisis: Other Concerns

We are a social species. Social distancing, therefore, is not a natural phenomenon for us. Quarantined, we are expected to avoid large gatherings and close contact with others. We are expected to suppress our 'evolutionarily hardwired impulses for connection'. It is not easy. But it is the requirement of the time. Several questions are raised regarding social distancing and its potential social and psychological impact. George Miller in Science Journal writes," Social distancing prevents infections, but it can have unintended consequences." Some of his observations are:

1. Social isolation can increase the risk of a variety of health problems.
2. Social contacts buffer the negative effects of stress. "Just knowing that you have someone you can count on if needed is enough to dampen some of those responses even if [that person is] not physically present."
3. Awareness of these issues will prompt people to stay connected and take positive action.
4. Ability to handle social isolation and stress vary among the individuals. "Someone who is already having problems with, say, social anxiety, de-

pression, loneliness, substance abuse, or other health problems is going to be particularly vulnerable."

5 The belief that war would eventually be won no matter how bad things would go, would ultimately prevail.

6. "Collective effervescence" (sharing emotional excitement with people) magnifies the sensation that that you're something larger than yourself.

7. Giving support is more beneficial than receiving it.

LANCET in one of its issues has reviewed the psychological impact of quarantine and has pointed out a few important issues.

1. After quarantine avoidance behaviours, both in patients and healthcare workers can be seen.
2. Stressors include longer quarantine duration, infection fears, frustration, boredom, inadequate supplies, inadequate information, financial loss, and stigma. Longer durations of quarantine were associated with poorer mental health, specifically, post-traumatic stress symptoms, avoidance behaviours, and anger.
3. Lack of clarity, lack of transparency about the risk, clear guidelines and quarantine protocols are some of the predictors of post-traumatic stress symptoms.
4. Financial problem and loss of work are key post-quarantine stressors.
5. Stigma of quarantine is an issue. Participants in several studies reported that others were treating them differently: avoiding them, withdrawing social invitations, treating them with fear and suspicion, and making critical comments. General education about quarantine helps in the reduction of stigma.
6. Depriving people of their liberty for the wider public good is often contentious and needs to be handled carefully.

The COVID crisis has exposed our fragility at several fronts. It concerns, besides health, how we treat our planet. Community life is experiencing a vigorous resurgence. The pandemic prompts us to make a few personal sacrifices intended to benefit the whole. Our collective behaviour must change. We are different when we are in a crowd than when we are alone. We have to move from "ignorance, hate, and fear" to "curiosity, compassion, and courage." If there is a need to maintain social distancing, there is also a need to open the heart and mind. Denial worsens the situation. Blame game doesn't help. Trust is essential for any healing. Don't feign blindness. Coordinate and collaborate. False propaganda is dangerous. Shift from ego to eco. We have noticed that meetings are not as important as we made them out to be. Many things can be done from home, and for that will and discipline are required. Take a relook at 'essentials' and 'non-essentials'. Be careful not to delete some 'non-essentials' from the list. When there is change, there is addition as well as subtraction. We have few challenges in front of us: planetary healing and societal renewal, reshaping learning and leadership structures, premortem and post mortem approaches in planning and execution. We need to build schools of transformation technologies with focus on not only to design, but also to build.

We must partly take the responsibility for the situation we are in. Outbreak of new infectious diseases is not new for us. In future we will face more such situations. This is because we are undergoing significant change in land use pattern. We are disturbing wild life habitat due to aggressive agriculture, forestry, mining and oil exploration activities. This pandemic will follow another one soon, unless we change our living pattern, and that includes trade in animal products. If we don't do that we can't stop 'jumping of viruses from animals to humans'.

How long the lockdown will continue? How long we will allow this weird virus to disturb our daily routine? How long this 'cytokine storm' will reside in our bodies and minds? This virus, being new, will take time to develop immunity, like any other virus. Are not precautions enough for preventions? "Good quality supportive care based on an understanding of the pathophysiology of the infection is what saves most lives." (19.)

Social media must play a more positive role. Don't create panic even if 'death' sells more than 'life'. Reduce irrationality. Don't take the responsibility of what is not your competence. Half-truths are worse than complete lies. Use scientific intelligence and rationale to deal with the pandemic. Desperation and the urge of self-promotion do not lead to solutions we are looking for. Hastier than thou is only good for the menace and not for the humanity. Too much pessimism makes one incurable. Optimum optimism helps, as it always keeps some space for pessimism. Arrival of 'magic bullet' will take time. 'New normal' will also take time to come.

It is true many predictions have gone wrong. We notice that some scientific advisers are getting death threats "for crippling the economy". Is it justified? Scientists are blamed for both 'too much lockdown' and 'too ready to relax restrictions'. Why don't we realize that good science requires time. Moreover, good science doesn't create Frankenstein.

Someone rightly said, "Let us remember that we are only as strong as the weakest health system in our interconnected world." Covid19 is not an equal-opportunity killer, writes Lizzie Wade (20). "In hard-hit New York City, Latino and black people have been twice as likely to die from COVID-19 as white people." Inequalities put people at higher risk. Often inequality worsens the crisis. People have the problem of denial. Please do not be in a denial mode, learn to live with the realities of the virus.

Should elderly be sacrificed because we don't need them anymore? "But this is not acceptable to many cultures including the Indian culture where we have respect for the elderly," is how responds Nassim Nicholas Taleb to a question related to the pandemic. "So people are noticing the two worlds. One is the ancient world imbued with ancient values and second is the new world that succeeded economically over the past 75 years where moral advancement is not there. That is going to cost them. ....This is a cost benefit analysis and a lot of countries including India, countries in Southeast Asia, Middle East and Mediterranean would refuse on whole grounds."

The post-Covid world has too many contradictory expectations from us. We want remote digital learning, and at the same time want to keep student mobility and interaction. We want 'open access publications' without causing much agony to 'blue sky funding'. Coronavirus is a naturally occurring pathogen. It doesn't need a visa to enter into any country. It has no restrictions to travel to any country. Its genetic structure is known. Its proteins are being characterised. It can be bioengineered. It can be used as a bioweapon. More than attacking the lives of its citizen, it can attack citizen's most prized possession, its industry, its economy. It can change warfare strategy. It can make humans morally more vulnerable. If we can get over the menace of a virus, we can live in a better world. We can create a new order. As they say, life science is too important to be left alone in the care of a few. It needs the care and supervision of many.Let us join hands in creating the new order. Perhaps that was long overdue.

Let me end my article in a positive note. I will state a few salient points highlighted by our Prime Minister Narendra Modi during his address to the Nation on May 12, 2020.

1. Time has taught us that we have to now think local. Even global brands were once local. We need to be vocal for our local products. Not just buy but also publicise them.

2. This crisis is big but we cannot allow COVID19 to define us.

3. "Sarvam Aatmam Vasham Sukham". That which is in your control gives you happiness. We have to move ahead with new energy. We and we will make India self-reliant.

4. Self-reliance is possible with self-confidence and acting with strength.
5. Today we have the will and the way.
6. Five pillars will prop up this edifice; 1) Economy that doesn't talk of incremental change but a quantum jump. 2) Infrastructure 3) Technology driven systems 4) Demography is our source of energy 5) Demand - we need to stoke demand and every spoke of our supply chain has to be strengthened. This supply chain will have the aroma of our country and the sweat of our labour.
7. Our definition of self-reliance is about including everyone, who believe the earth as mother. When such a country becomes self-reliant, then the prosperity and happiness of the world is included in it. The world is convinced that we can contribute for the betterment of humankind.
8. At this time of crisis we have to be more resolute. Our resoluteness should be bigger than the crisis. It is our responsibility to look at this crisis to make this India's century. The way for this is only one: A self-reliant India.

Jai Hind

Notes

1. Mindset Defines the Contours of Life

1. Purnendu Ghosh, Neural suitcase tells the tales of many minds, Partridge, 2014.
2. Ferris Jabr, Why we need to study the brain's evolution in order to understand the modern mind, Scientific American, September 20, 2012.
3. David DeSteno, The truth about trust: How it determines success in life, love, learning, and more, Hudson Street Press, an imprint of Penguin Group, 2014.
4. Daniel Kahneman, Jonathan Renshon, Why Hawks Win, Foreign Policy, October 13, 2009.
5. Daniel Levitin, This is Your Brain on Music, Plume, 2007.
6. Nicholas Carr, It doesn't matter, Harvard Business Review, May 2003.
7. Ellen J Langer, Mindfulness, Da Capo Press, 1989.
8. Carol S. Dweck, Mindset: The New Psychology of Success, Random House, 2006.
9. Saul McLeod, Pavlov's Dogs, Simply Psychology, 2018.
10. Phil McKinney, Why Questions Matter, The world financial review, September 16, 2012.
11. Robin Maratz Henig, Darwin's God, The New York Times Magazine, March 4, 2007.
12. Leon Lederman, The God Particle, https://faculty.washington.edu/lynnhank/Lederman.pdf
13. Richard Nisbett, The geography of thought: How Asians and Westerners think differently...and why, Free Press, 2004.
14. Bipin Chandra Pal, The soul of India, Rupa Publications, 1900.

15. Jai B.P. Sinha, Psycho-social analysis of the Indian mindset, Springer, 2014.
16. Jonah Berger, Invisible influence: The hidden forces that shape behavior, Simon & Schuster, 2017.
17. Will Durant, The case for India (1931), re-published by Gyan Publishing House, 2017.
18. Jean Drèze , Amartya Sen, An uncertain glory: India and its contradictions, Allen Lane, 2013.
19. Amartya Sen, The Argumentative Indian: Writings on Indian History, Culture and Identity, Penguin, 2006.
20. Pankaj Mishra, India in Mind, Picador, 2005.
21. Pavan K. Varma, Being Indian: Inside the Real India, William Heinemann Ltd, 2005.
22. Caleb Scharf, Where do minds belong, aeon, 22 March, 2016.
23. Michio Kaku, The future of the mind: The scientific quest to understand, enhance, and empower the mind, Doubleday, Feb 2014.
24. David Gelernter, Dream-Logic, THE Internet and asrtificial thought, EDGE, 7 July, 2010.
25. Alan Jasanoff, We are more than our brains, https://attentiontotheunseen.com/2018/07/31/we-are-more-than-our-brains/

2. Scince Refines the Spirit of Life

1. Karl Popper , The logic of scientific discovery, Routledge, 2002.
2. Carl Sagan, The demon-haunted world: Science as a candle in the dark, Ballantine Books, 1997.
3. US National Academies of Sciences, Engineering, and Medicine, Trust and Confidence at the Interfaces of the Life Sciences and Society: Does the Public Trust Science? A Workshop Summary. Washington, DC: The National Academies Press, 2015.
4. Evelyn Fox Keller, A feeling for the organism, Times Books, 1984.
5. David Galenson , Clayne Pope, Collaboration in Science and Art, The Huffington Post, 23 July, 2012.
6. Dean Keith Simonton, Creativity in science – Chance, Logic, Genius and Zeitgest, Cambridge University Press, 2004.

7. Bruce Alberts, Designing scientific meetings, Editorial, Science, Vol. 339, 15 Feb. 2013.
8. The Roundtable on Public Interfaces of the Life Sciences (PILS) of the National Academies of Sciences, Engineering, and Medicine, a workshop in Washington, DC on Public Trust in Science, May 5-6, 2015.
9. Freeman Dyson, Progress in Religion, https://www.edge.org/conversation/freeman_dyson-progress-in-religion.
10. Paul Davies, Taking science on faith, The New York Times, Nov. 24, 2007.
11. William Phillips, Science, Faith, and the Nobel Prize, http://ps100.byu.edu/phillips%20forum%20transcription.pdf
12. Gary Klein, Performing a project premortem, Harvard Business Review, SEPTEMBER 2007.
13. Daniel Kahneman's favorite approach for making better decisions, Farnam Street, https://fs.blog/2014/01/kahneman-better-decisions/
14. Lucie Laplanea, Paolo Mantovanic, Ralph Adolphsd, Hasok Change, Alberto Mantovanif, Margaret McFall-Ngaih, Carlo Rovellii, Elliott Soberj, and Thomas Pradeua, PNAS, vol. 116, no. 10, 3948–3952, March 5, 2019.
15. David Gelernter, The Closing of the Scientific Mind, Commentary Magazine, JAN. 1, 2014.
16. Thomas Kuhn, The structure of scientific revolutions, University of Chicago Press, 1996.
17. Mary Catherine Bateson, Composing a further life: The age of active wisdom, Vintage, 2011.
18. Douglas Rushkoff, Program or be programmed: Ten commands for a digital age, Soft Skull Press, 2011.
19. John C. Polanyi, On Being a Scientist: A Personal View, The Globe and Mail (Canada), 29 April 2000.

3. Engineering Designs the Vision of Life

1. The Engineer of 2020: Visions of Engineering in the New Century http://www.nap.edu/catalog/10999.html
2. Design innovation at Berkeley Engineering, 2014 https://engineering.berkeley.edu/2014/08/design-innovation-berkeley-engineering

3. Andrea Bandelli, Being creative to continue meaningful lives, Medium, March 24, 2017.
4. Kieren Egan, The educated mind: How cognitive tools shape our understanding, University of Chicago Press, 1997.
5. Lisa R Lattuca, Patrick T Terenzini, J Fredfericks Volkswein, George D Peterson, The changing face of engineering education, The Bridge, December 3, 2008.
6. G Wayne Clough, Reforming Engineering Education (editorial), THE BRIDGE, DECEMBER 3, 2008.
7. W A Wulf, G M C Fisher, Issues in science and technology, Vol 18 No 3 Spring 2002.
8. Phil Kaminsky, Berkeley Engineer, November 1, 2017.
9. Howard Gardner, Truth, Beauty, and Goodness: Education for All Human Beings, Edge (www.edge.org), 20 September, 1997.
10. National Academies of Sciences, Engineering, and Medicine, The integration of the humanities and arts with sciences, engineering, and medicine in higher education: Branches from the same tree. Washington, DC: The National Academies Press, 2018.
11. National Academies of Sciences, Engineering, and Medicine, A Vision for the Future of Center-Based Multidisciplinary Engineering Research: Proceedings of a Symposium, Washington, DC: The National Academies Press. 2016.
12. India Today, July 13, 2016 UPDATED: February 6, 2018).
13. B Michael Aucoin, From engineer to manager – Mastering the transition, Artech House Technology Management and Professional Development Library, 2002.
14. Carol Milano, Innovation and research: The human factor, Science, Sep. 16, 2011.
15. Tony Wagner, Creating innovators: The making of young people who will change the world, Scribner, 2015.
16. Ruth Graham, The global state of the art in engineering education, neet.mit.edu, March 2018.
17. W I B Beveridge, The art of scientific investigation, Edizioni Savine, 2017.

18. Gary Marcus, Unknown Unknowns, Edge, https://www.edge.org/response-detail/23691

19. R Keith Sawyer, Explaining Creativity: The Science of Human Innovation, Oxford University Press, 2006.

20. Dean Simonton, Creativity in Science: Chance, Logic, Genius, and Zeitgeist, Cambridge University Press, 2004.

21. Joichi Ito, Design and Science, Jan 12, 2016, https://jods.mitpress.mit.edu/pub/designandscience/branch/2/0

22. Rex Eugene Jung, Richard J Haier, Creativity and intelligence: Brain networks that link and differentiate the expression of genius, In book: Neuroscience of Creativity, pp.233-254, 2013.

23. Bernard Marr, www.forbes.com Sep 2, 2018.

24. Cornelius Baur, Dominik Wee, Manufacturing's next act, McKinsey & Company, June 2015.

25. Martin, https://www.cleverism.com/industry-4-0/)

26. https://www.proschoolonline.com/blog/what-is-industry-4-0-and-is-india-prepared-for-the-change)

27. Vinayak Dalmia, Kavi Sharma, World Economic Forum, 13 February, 2017.

28. Navi Radjou, Jaideep Prabhu, Simone Ahuja, Harvard Business Review, JULY 02, 2012.

29. Nirmalya Kumar, Phanish Puranam, Frugal engineering: An emerging innovation paradigm, IVEY Business Journal, March / April 2012.

30. Geoffrey West, Why cities keep growing, corporations and people always die, and life gets faster, A Conversation, Edge, 23 May, 2011.

31. Ram Mudambi, Haritha Saranga, Andreas Schotter, Mastering the Make-in-India Challenge, MIT Sloan Management Review Magazine June 08, 2017.

32. Bruce Alberts, Why National Science Academies, Science, Vol. 339, Issue 6123, pp. 1011, 01 Mar 2013.

33. George Constable, Bob Somerville, A century of innovation: Twenty engineering achievements that transformed our lives. Washington, D.C., National Academies Press, 2003.

34. http://www.nap.edu/catalog/10999.html)

35. Prem Krishna, K V Raghavan, S S Chakraborty, Purnendu Ghosh, Vision, Mission and Values INAE 2037, https://www.inae.in/core/assets/fortuna-child/img/INAEVisionmissionandvalues.pdf

4. Convergence of Engineering and the Science of Life

1. Erwin Schrödinger, What is life, Cambridge University Press; Reprint edition, 2012.

2. Anna Deplazes-Zemp, Nikola Biller-Andorno, Explaining life - Synthetic biology and non-scientific understandings of life, EMBO Rep. ,13(11): 959–963, 2012.

3. Drew Endy, Foundations for engineering biology, Nature, 438, 449, 2005.

4. Convergence: Facilitating transdisciplinary integration of life sciences, physical sciences, engineering, and beyond (2014) https://doi.org/10.17226/18722

5. Peter A Carr, George M Church, Genome engineering, Nature Biotechnology 27, 1151, 2009.

6. Mario Capecchi, Martin Evans, Oliver Smithies, The Nobel Prize in Physiology or Medicine, 2007.

7. Seirian Sumner, A synthetic world, https://www.edge.org/response-detail/23782

8. Stelios Papadopoulos, Business Models in Biotech, Nature Biotechnology 18, IT3 - IT4, 2000.

9. Guido Lanza, Building today's platform company, Nature Biotech Bioentrepreneur Published online: 30 June 2009.

10. Irene Berner, Susan Dexter, Parrish Galliher, Global evolution of biomanufacturing, Bioprocess International, March 1, 2013.

11. J D Keasling, Manufacturing molecules through metabolic engineering, Science 330, 1355, 2010.

12. National Academies of Sciences, Engineering, and Medicine, Preparing for Future Products of Biotechnology. Washington, DC: The National Academies Press, 2017, https://doi.org/10.17226/24605)

13. Alan Walton, Some Thoughts on Bioentrepreneurship, http://www.nature.com/bioent/2003/030101/full/nbt0598supp_7.html

14. Emily Mullin, MIT TECHNOLOGY REVIEW, March/April 2017 Issue, https://www.technologyreview.com/s/603498/10-breakthrough-technologies-2017-gene-therapy-20/)
15. Steve Connor, MIT TECHNOLOGY REVIEW March/April 2017 Issue, https://www.technologyreview.com/s/603499/10-breakthrough-technologies-2017-the-cell-atlas/)
16. https://healthmatters.nyp.org/absorb-stent/)
17. Jihun Park, Joohee Kim, So-Yun Kim, Woon Hyung Cheong, Jiuk Jang, Young-Geun Park, Kyungmin Na, Yun-Tae Kim, Jun Hyuk Heo, Chang Young Lee, Jung Heon Lee, Franklin Bien, Jang-Ung Park, Science Advances, Vol. 4, no. 1, eaap9841, 24 Jan 2018.
18. https://www.iflscience.com/health-and-medicine/new-saliva-test-may-be-able-detect-cancer-just-10-minutes/

5. Technology Revolution And Optimism

1. Matt Ridley, The Rational Optimist: How Prosperity Evolves, Harper Perennial; Reprint edition, 2011.
2. Jennifer Hochschild, Alex Crabill, Maya Sen, Technology optimism or pessimism: How trust in science shapes policy attitudes toward genomic science, Issues in technology innovation, December 2012.
3. Kevin Kelly, What technology wants, Penguin Books, 2011.
4. Jason Pontin, Three Commandments for technology optimists, IDEAS, 10 October, 2018.
5. Jared Diamond, Guns, Germs And Steel, RHUK, 1998
6. Alison Wolf, Does education matter? Myths about education and economic growth, Penguin, 2002.
7. W F Durand, The engineer and civilization, Science, Vol. 62, Issue 1615, pp. 525-533, 11 Dec 1925.
8. Dugald C Jackson, Engineering in our early history, Proceedings Symposium on the early history of science and learning in America, February 13, 1942.
9. Freeman Dyson, Is science mostly driven by ideas or by tools, Science, 338, no. 6113, 1426-27, 2012.
12. Dean Keith Simonton Creativity in Science: Chance, Logic, Genius, and Zeitgeist, Cambridge University Press, 2004.

13. Yuval Noah Harari, Homo Deus – A brief history of tomorrow, Vintage, 2017.
14. Robert Pollack, Rethinking our vision of success, Edge (www.edge.org) 2019
15. Sadhguru, The inner Engineering: A Yogi's guide to joy, Penguin, 2016.
16. Matt Ridley, The Evolution of Everything, Harper Collins, 2015.

6. Future of Work

1. Daniel Kahneman, Thinking, Fast and Slow, Penguin Press, 2012.
2. Mihaly Csikszentmihalyi, Creativity: The Psychology of Discovery and Invention, Harper Perennial Modern Classics, 2013.
3. Howard Gardner, Editor, Good Work: Theory and Practice, http://thegoodproject.org/pdf/GoodWork-Theory_and_Practice-with_covers.pdf
4. Susan Verducci, In God We Trust; All Others Bring Data" Trust and Accountability, in GoodWork, http://thegoodproject.org/pdf/GoodWork-Theory_and_Practice-with_covers.pdf
5. Niklas Luhmann, Familiarity, Confidence, Trust: Problems and Perspectives, In Diego Gambetta (ed.), Trust: Making and Breaking Cooperative Relations. Blackwell, 1988.
6. Peter M Senge, The Fifth Discipline-The art and practice of the learning organization, RHUK, 2006.
7. Jeffrey Pfeffer, The Human Equation, Harvard Business School Press,1998.
8. Barbara Fredrickson, Positivity, Harmony, 2009.
9. Warren Bennis, An Invented Life: Reflections On Leadership And Change, Basic Books, 1993.
10. Abraham Zaleznick, Managers and leaders: Are they different, Harvard Business Review 1977.
11. Jim Collins, Good to Great: Why Some Companies Make the Leap … and Others Don't, Harper Business, New York, 2001.
12. James Manyika, Technology, jobs, and the future of work, https://www.mckinsey.com/~/media/McKinsey/Featured%20Insights/Employment%20and%20Growth/Technology% 20jobs%20and%

20the%20future%20of%20work/MGI-Future-of-Work-Briefing-note-May-2017.ashx

13. The State of the Debate, a report prepared by the Roosevelt Institute for the Open Society Foundations, April 2015.
14. Claudia Goldin , Lawrence F. Goldin, Larry Katz, The Race Between Education and Technology, Belknap Press: An Imprint of Harvard University Press; 2010.
15. Samir Saran, Vivan Sharan, Occasional Papers Mar 21 2018.
16. Shoshana Zuboff, In the Age of the Smart Machine: The Future of Work And Power, Basic Books; Reprint edition, 1989.
17. Evgeny Morozov, The Net Delusion, Penguin, 2012.
18. Paul Saffo, Technology Forecaster at Stanford University
19. Yuval Noah Harari, Homo Deus – A brief history of tomorrow, Vintage, 2017.
20. Jonathan Gershuny and Oriel Sullivan, What We Really Do All Day: Insights from the Centre for Time Use Research, Pelican, 2019.

7. COVID Crisis and Technology Update

1. Kristian G. Andersen, Andrew Rambaut, W. Ian Lipkin, Edward C. Holmes & Robert F. Garry, The proximal origin of SARS-CoV-2, Nature Medicine, 26, 450, 2020.
2. Mark Terry, International Researchers Conclude COVID-19 is Not Man-Made, BioSpace, Mar 18, 2020.
3. Michele Wucker, Washington Post, March 18, 2020.
4. James Pethokoukis, The Week, March 8, 2020.
5. David Cyranosk, Nature, 4 MAY, 2020.
6. Megan Scudellari, The sprint to solve coronavirus protein structures — and disarm them with drugs, Nature, 15 May, 2020.
7. NidhiSubbarama, Coronavirus tests: researchers chase new diagnostics to fight the pandemic, Nature, 23 March, 2020.
8. Ewen Callaway, Coronavirus vaccines, Nature, vol 580, 30 April, 2020.
9. (http://timesofindia.indiatimes.com/articleshow/75770647.cms?utm_source=contentofinterest&utm_medium=text&utm_campaign=cppst).

10. Heidi Ledford, Dozens of coronavirus drugs are in development — what happens next?, Nature News, 14 May, 2020.
11. Ewen Callaway, Nature, April 30, 2020.
12. RhythmaKaul, Hindustan Times, May 10, 2020.
13. Timothy Devinney and Grahame Dowling, THE, May 14, 2020.
14. (https://a16z.com/2020/04/18/its-time-to-build/).
15. David Rotman, MIT Technology Review, April 25, 2020.
16. John Robinson, The Manufacturer, 2 April, 2020.
17. Greg Petro,FORBES, March 6, 2020.
18. Politico Magazine, 19 March, 2020).
19. HimadriBarthakur and Sumit Roy, COVID 19: it is time to take a step back from stories of 'Doom and Gloom', WIRE, May 12, 2020.
20. Lizzie Wade, An unequal blow, Science, 15 May 2020, 2020.

Printed in the United States
by Baker & Taylor Publisher Services